WIZARDS
ALIENS
AND
STARSHIPS

WIZARDS ALIENS

AND STARSHIPS

PHYSICS AND MATH IN FANTASY AND SCIENCE FICTION

CHARLES L. ADLER

PRINCETON UNIVERSITY PRESS • PRINCETON AND OXFORD

Jacket Illustration: Chesley Bonestell, *Space Station, Ferry Rocket, and Space Telescope 1,075 Miles above Central America* (1952). Reproduced courtesy of Bonestell LLC.

Library of Congress Cataloging-in-Publication Data

Adler, Charles L.
    Wizards, aliens, and starships : physics and math in fantasy and science fiction / Charles L. Adler.
        pages cm
    Summary: "From teleportation and space elevators to alien contact and interstellar travel, science fiction and fantasy writers have come up with some brilliant and innovative ideas. Yet how plausible are these ideas—for instance, could Mr. Weasley's flying car in Harry Potter really exist? Which concepts might actually happen—and which ones wouldn't work at all? *Wizards, Aliens, and Starships* delves into the most extraordinary details in science fiction and fantasy—such as time warps, shape changing, rocket launches, and illumination by floating candle—and shows readers the physics and math behind the phenomena. With simple mathematical models, and in most cases using no more than high school algebra, Charles Adler ranges across a plethora of remarkable imaginings, from the works of Ursula K. Le Guin to *Star Trek* and *Avatar*, to explore what might become reality. Adler explains why fantasy in the Harry Potter and Dresden Files novels cannot adhere strictly to scientific laws, and when magic might make scientific sense in the muggle world. He examines space travel and wonders why it isn't cheaper and more common today. Adler also discusses exoplanets and how the search for alien life has shifted from radio communications to space-based telescopes. He concludes by investigating the future survival of humanity and other intelligent races. Throughout, he cites an abundance of science fiction and fantasy authors, and includes concise descriptions of stories as well as a glossary of science terms. *Wizards, Aliens, and Starships* will speak to anyone wanting to know about the correct—and incorrect—science of science fiction and fantasy"— Provided by publisher.
    Includes bibliographical references and index.
    ISBN 978-0-691-14715-4 (hardback : acid-free paper)
    1. Fantasy literature—History and criticism.   2. Science fiction—History and criticism.
    3. Physics in literature.   4. Mathematics in literature.   5. Physics—Miscellanea.
    6. Mathematics—Miscellanea.   I. Title.

PN3433.8.A35 2014
    809.3'8762—dc23   2013027794

British Library Cataloging-in-Publication Data is available

This book has been composed in Minion Pro and League Gothic

Printed on acid-free paper. ∞

Typeset by S R Nova Pvt Ltd, Bangalore, India
Printed in the United States of America

10 9 8 7 6 5 4 3 2 1

To Poul Anderson, who wrote it better, shorter, and earlier

CONTENTS

WIZARDS
ALIENS
AND
STARSHIPS

CHAPTER ONE

PLAYING THE GAME

Dear Roger,

(XXX) and I have been exchanging letters for some time. As a fan, he's strange; he likes the science better than the fiction. Wants me to quit futzing with the plot and characters and get on with the strange environments. He plays The Game: finds the holes in the science and writes in. I like him. . . .

—LETTER FROM LARRY NIVEN TO ROGER ZELAZNY, JANUARY 3, 1974

1.1 THE PURPOSE OF THE BOOK

When I was young, back in the 1970s and 80s, I read a lot of science fiction. I read a lot of other stuff, as well, but science fiction (and fantasy) filled a need that other literature simply didn't. I tended to read "hard" science fiction, that is, stories plotted around hard science: physics, astrophysics, giant engineering projects, and the like. The worlds these stories portrayed, where space travel was common, human problems such as poverty were nearly eliminated, and conflicts centered on larger-than-life issues, always seemed to me more compelling than

1

human dramas that revolved around why someone didn't love someone else.

My tastes have changed since then, but the initial thrill of these stories has never really left me. I am a scientist because of my initial love of these tales. A chill still runs down my spine whenever I look at a Hubble Telescope photo or learn of a new exoplanet discovered. I live in hope that I will be alive when life on other planets is discovered. I still want to take a vacation to the Moon or to an orbiting satellite. These thrills are tempered by my adult realization that much of what goes into science fiction is quite unrealistic. This book is written for my fifteen-year-old self, and other readers like him, who would like to know which parts of science fiction are based on real science, and therefore in some way plausible, and which parts are unrealistic. This is the book I would have wanted to read when I was young. Just as for Niven's correspondent, my interest in science fiction was mostly in the strange environments, the new worlds, the alien life, the superscience it portrayed. I wanted to know which parts were (potentially) real and which weren't. To a large extent, that is why I eventually became a physicist.

Almost any science fiction story has a lot of incorrect science. This doesn't make the story bad or invalid. Some authors, like Larry Niven, are almost obsessive in trying to get the science right; most are more lackadaisical about it. However, the standards for the profession are pretty high: no science fiction writer can be really esteemed accomplished unless he or she has a thorough knowledge of basic physics, chemistry, biology, astrophysics, history (ancient and modern), sociology, and military tactics; and besides all this, must possess a certain something in their air and manner of writing, or their profession will be but half-deserved. (Improvement of their minds by extensive reading goes without saying.) Science fiction writers do not have the same opportunities as research scientists do to stay up-to-date in their research fields, and writing science fiction involves a lot more fields than most research scientists can keep up with.

This book is one physicist's attempt to discuss the science, particularly the physics and mathematics, that goes into writing hard science fiction. As an added bonus, I also take a look at physics in fantasy writing: there's more in it than meets the eye. This is not an attempt to predict the future: as G. K. Chesterton pointed out, most of the fun in predicting the future comes from burying the people who attempt to do it [50]. Rather, I stick to the science used in crafting the stories. There are many books dedicated to the literary criticism of science fiction; this book is devoted to its *scientific* critique. As such, my choice of which literature to use is dictated both by my own reading and by the needs of the book. I tend to avoid writers who don't make much use of science in their stories, except occasionally to comment on their errors. I also tend to stick to literature, that is, novels and short stories, although I occasionally comment on science fiction movies or television shows as well.

Many have gone down this path before me, scientists and writers alike (and a few who were both). The preeminent standout among science fiction writers is Poul Anderson, to whom this book is dedicated, for his essays "How to Build a Planet" and "On Thud and Blunder." I read both when I was a teen; this book would not have been written but for his example. Isaac Asimov and Arthur C. Clarke both wrote many essays on science. Larry Niven has written several essays on the scientific aspects of teleportation, time travel, and other science fiction themes. Almost from the beginning of the modern era, scientists have written essays on science fictiony ideas, and I reference them where appropriate. This book is mainly synthetic rather than original, although I think there are a few new things in it, such as the discussion of candlelight in the Harry Potter series in chapter 3.

1.2   THE ASSUMPTIONS I MAKE

David Gerrold has written that science fiction authors by necessity almost always involve bits in their work that defy the laws of science as we know them. He refers to places where this happens as instances of "baloneyum." His advice is that beginning authors limit themselves to only one piece of baloneyum per story, experienced authors perhaps as many as two, and only grandmasters put in three instances [94]. It's a good rule.

In this book I have followed a similarly conservative path. In analyzing science fiction my assumptions are that the laws of physics as we understand them now are pretty much correct. They are *incomplete*;

we don't know all of them, but the incompleteness doesn't really affect most science fiction stories. In particular, I assume that Newton's laws of motion are good enough to describe things larger than atoms, that Einstein's theory of relativity is correct, and that quantum mechanics is the correct description of nature on the microscopic scale. The one example of baloneyum I indulge in is in the consideration of faster-than-light travel and, equivalently, time travel, which appear to be impossible from almost everything we know about physics—but perhaps not quite.

1.3 ORGANIZATION

There are four large sections of the book. Each of them contains several chapters centered on a given theme. The sections are:

1. "Potter Physics." This first section explores physics as used and abused in fantasy novels and series. I've chosen two examples of "urban fantasy" novels to focus on, the Harry Potter series by J. K. Rowling and the Dresden Files novels by Jim Butcher. The issues here are different from those in the rest of the book, for fantasy, by its very nature, cannot adhere strictly to scientific laws. However, we can ask whether the series are at least internally consistent, and whether the magic used in the series makes sense when in contact with the muggle world.

2. "Space Travel." This is the largest single section, consisting of nine chapters, as befits the subject. Space travel is perhaps *the* theme of science fiction, to the point that it almost defined the genre from the 1930s to the 1980s. One goal of this section is to examine not only the scientific issues involved in space travel but also the economic ones. The big question is, why isn't space travel cheap and common now, as it was certainly foretold to be in almost all of "golden age"[1] science fiction?

3. "Worlds and Aliens." This section consists of four chapters exploring the other major theme of science fiction, the possibility of life on other worlds in our Solar System and elsewhere.

4. "Year Googol." This part explores the potential survival of humanity (or other intelligent species) into the far distant future, along lines laid down originally by the writer Olaf Stapledon and the physicist Freeman Dyson.

My choice of subject matter, like the organization of the book, is idiosyncratic. The book is a loose collection of essays more than a unified text. I write about those aspects of science fiction and fantasy that most interest me. My hope is that my readers are similarly interested. By necessity, I concentrate on those writers whom I know the best, meaning American and British science fiction writers. Since I know the "golden age," New Wave, and early cyberpunk literature best, this may give the book an antiquated feel. I try to include ample description of these stories so that anyone reading the book can understand the scientific points I am trying to make.

A set of problems has been prepared for instructors intending to use this book as a class text. For space reasons, we have placed these problems, organized by chapter, on a website (press.princeton .edu/titles/10070.html). I've also included solutions and hints for the problems. This book cannot be used to replace a physics textbook, but it could be used in a specialized course.

1.4 THE MATHEMATICS AND PHYSICS YOU NEED

I expect the readers of this book to be able to read and use algebraic equations, and to understand them on some level. I intend the book as a working book for science fiction enthusiasts who have at least a decent knowledge of algebra and know what calculus means.

The equations I introduce don't exist in a vacuum; they are mostly drawn from physics, and represent physical quantities. That is, unlike pure mathematics, there is always some connection with the real (or, at least, science fiction) world that is expressed by them. In most cases I explain the equations in detail but do not derive them from basic principles. This is unlike what happens in most physics courses, where the emphasis is as much on deriving the equations as on using them. Since most of my readers aren't physicists, I will explain how the equations are used and why they make sense. I also want my readers to have a conceptual understanding of calculus. There are only a few places where this will crop up, so it isn't essential, but it is useful to know what is meant when I use the terms "derivative" and "integral."

Physics is the science central to this book. Appendix 1 at the end of the book reviews Newton's laws of motion, which are central to any understanding of physics. Just as a knowledge of grammar and spelling is needed for reading and writing, a knowledge of Newton's laws is needed for any understanding of physics. Newton's laws describe how things move on the macroscopic scale; that is, they are a good description of things larger than atomic size. However, they are only approximations to the truth. The laws of quantum mechanics are the real way things work. It is characteristic of physics that the underlying laws are difficult to see directly. Why this is so, and why Newton's laws are good approximations to the true fundamental laws of nature, are questions beyond the scope of this book to answer. If readers are interested in this, there are dozens of good books that examine these questions. I strongly recommend two books by Richard Feynman, *The Character of Physical Law* and (for those who have a physics background) *The Feynman Lectures on Physics*, particularly book 2, chapter 20 [81][85].

1.5 ENERGY AND POWER

Energy and power, which is the rate at which energy is converted from one form to another, are the key points to understand for this book. Energy is useful because it is conserved: it can be transformed from one form to another, but not created or destroyed. I use energy conservation, either implicitly or explicitly, in almost every chapter of the book. A few of the forms that energy can take are the following:

- Gravitational potential energy: This is the energy that pairs of objects possess by virtue of the gravitational attraction between the members of the pair. This form of energy is very important for any discussion of space travel.
- Chemical potential energy: This is the energy resulting from the spacing, composition, and shape of chemical bonds within a molecule. Chemical reactions involve changes in these properties, which usually means changes in chemical potential energy. In an exothermic reaction, the chemical potential energy is less after the reaction than before it. Energy is

"released" during the course of the reaction, usually in the form of heat. An endothermic reaction is the opposite: energy must be added to the reaction to make it proceed.

- Nuclear energy: This is the energy resulting from the structure and composition of the atomic nucleus, the part of the atom containing the protons and neutrons. Transformations of nuclei either require or release energy in the same way that chemical reactions do, except on an energy scale about one million times higher.

- Mass: Mass is a form of energy. The amount of energy equivalent to mass is given by Einstein's famous formula

$$E = Mc^2, \tag{1.1}$$

where E is the energy content of mass M and c is a constant, the speed of light (3×10^8 m/s, in metric units). This is the ultimate amount of energy available from any form of mass.

- Kinetic energy: This is the energy of motion. Newton's formula for kinetic energy is

$$K = \frac{1}{2} Mv^2, \tag{1.2}$$

where K is the kinetic energy, M is the mass of the object, and v is its speed. This formula doesn't take relativity into account, but it is good enough for speeds less than about 10% the speed of light. If an object slows down, it loses kinetic energy, and this energy must be turned into another form. If it speeds up, energy must be converted from some other form into kinetic energy.

- Heat: Heat is energy resulting from the random motion of the atoms or molecules making up any object. In a gas at room temperature, this is the kinetic energy of the gas molecules as they move every which way, plus the energy resulting from their rotation as they spin about their centers. For solids or liquids, the energy picture is more complicated, but we won't get into that in this book.

- Radiation: Light, in other words. Light carries both energy and force, although the force is almost immeasurable under most circumstances. Most light is invisible to the eye, as it is at wavelengths that the eye is insensitive to.

In the units most often used in this book, energy is expressed in joules (J). The joule is the unit of energy used in the metric system. To get a feel for what a joule means, take a liter water bottle in your hand. Raise it 10 cm in the air (about 4 inches). You have just increased the potential energy of the water bottle by 1 J.[2] Other units are also used; in particular, the food calorie, or kilocalorie (kcal), will be used in several chapters. The kilocalorie is the amount of energy required to increase the temperature of 1 kg of water by 1°C (Celsius). It is equivalent to 4,190 J. Other units are defined as they come up in the chapter discussions.

Power is the rate at which energy is transformed from one form to another. The unit of power is the watt (W), which is 1 J transformed per second from one form of energy to another form. For example, if we have a 60 W light bulb, 60 J is being transformed *from* the kinetic energy of electrons moving through the tungsten filament in the light bulb *into* radiation, every second.

The different forms of energy and their transformations are the most important things you need to know to read this book. With this brief introduction to the subject, we are ready to start.

NOTES

1. Science fiction readers and critics divide up science fiction of the last century into different subgenres, which typically also follow one another chronologically. For example, the "golden age" covers the period from the end of World War II through the mid-1960s, when authors such as Robert Heinlein, Isaac Asimov, and Arthur Clarke were at the peak of their popularity. The major science fiction themes of this time period are space travel and alien contact. I also refer to the New Wave writers of the 1960s and 1970s, such as Brian Aldiss, and cyberpunk literature from the 1980s and later. Of course, many authors, such as Philip K. Dick, resist easy classification. Fantasy is similarly divided, into "sword and sorcery," "magic realism," and "urban fantasy," among other subgenres. Urban fantasies are very useful for this book because they allow side-by-side comparison of fantasy worlds with the real world in which the laws of physics hold sway.

2. To be specific, you have increased the potential energy of the pair of objects, the Earth plus the water bottle, by 1 J. Potential energy is always the property of a system, that is, of two objects or more, not of an individual object. However, because the Earth essentially hasn't moved, owing to its high mass, we typically talk about the lighter object of the pair as the one whose potential energy changes. Being specific about this is important only when the masses of the two objects in question are about equal.

PART I

POTTER PHYSICS

CHAPTER TWO

HARRY POTTER AND THE GREAT CONSERVATION LAWS

2.1 THE TAXONOMY OF FANTASY

The "physics of fantasy" seems like an oxymoron: by definition, fantasy doesn't concern itself with science but with magic. However, a lot of fantasy writers follow in the tradition of science fiction writers in trying to set up consistent rules by which their fantasy worlds operate. This is because many fantasy writers are science fiction writers as well. It is almost a universal trait: those who write quasi-realistic science fiction will also write quasi-realistic, rules-based fantasy; those who don't generally won't set up rules by which magic works.

Among the former is Ursula K. Le Guin, whose *Earthsea* trilogy has long descriptions of the "rule of names" underlying all magic. Her books include several lectures by magicians on exactly how this works. Many writers have found her works compelling enough to copy her rules in their own stories. Others base their magic rules on outdated scientific or philosophical ideas, as Heinlein did in his novella, *Magic, Incorporated*. Magicians in that book use the "laws" of similarity and so forth, to perform their magic. Randall Garrett in his Lord Darcy stories writes of a world in which magic (following these laws) has developed instead of science. The stories are full of descriptions of how the magic works and is used in solving crimes.

The popular writer J. K. Rowling in her Harry Potter novels does not attempt to have the magic in her books follow any known laws of science. Please don't misunderstand me: I love her books, but not for any attempt on her part to be consistent in how magic works. This is why this section of the book is called "Potter Physics": her body of work contains innumerable examples of magic being used in ways that violate physical law and are also internally inconsistent. She belongs solidly to the second class of writers.

These rules of magic don't have to follow known laws of physics—indeed, they can't, or else we'd call them science fiction instead of fantasy. The laws that many rules-based fantasy writers choose to keep are typically the most fundamental of the physical laws: the great conservation principles—the conservation of mass, energy, and momentum—and the second law of thermodynamics.

2.2   TRANSFIGURATION AND THE CONSERVATION OF MASS

> Harry spun around. Professor Moody was limping down the marble staircase. His wand was out and it was pointing at a pure white ferret, which was shivering on the stone-flagged floor, exactly where Malfoy had been standing.
>
> —J. K. ROWLING, *HARRY POTTER AND THE GOBLET OF FIRE*

The issue of shape-changing in the world of the Harry Potter novels is vexing. Since the 1800s one of the principal ideas of science has been the conservation of mass: the total mass in a closed system cannot change. In turning Draco Malfoy into a ferret, what did Professor Moody do with the rest of his mass? If we assume that Draco at age fifteen had a mass of about 60 kg and a small ferret has a mass of about 2 kg, where did he stash the other 58 kg? This is an issue for most fantasy writers when dealing with shape-changers. In *Swan Lake*, when Odette is changed into a swan by the evil wizard Rothbart, what happened to the rest of her? One can imagine some sort of weird biological process by which flesh morphs from one animal form to another, but where does the excess go?

A number of fantasy writers have dealt with this issue head-on. In Poul Anderson's *Operation Chaos*, the hero is a werewolf who explicitly states in the course of the book that his mass is the same in both states. This is OK for a werewolf, as the average adult wolf has about the same weight as a very light adult male, but it raises problems later on when the hero meets a weretiger. In human form, the magic user must maintain a weight of nearly 400 pounds simply to make a fairly small tiger. This entails health problems and severe psychological stress. In Niven's story "What Good Is a Glass Dagger?," the hero implicitly invokes the principle of conservation of mass when he says that he doesn't look too overweight as a human, but as a wolf he'd look ten years' pregnant.

This doesn't bother most fantasy writers, perhaps because it would impose too strong constraints on many fantasy stories if writers stuck to the conservation of mass. For example, when trapped by the sword-wielding barbarian, the beautiful sorceress can't turn herself into a dove and fly away. Instead, she'd have to turn into an ostrich and kick him in the ribs. (Actually, that would make a good story.) Thinking about where the mass goes leads to headaches. Einstein tells us mass is convertible into energy at a "cost" of 9×10^{16} J/kg. If somehow the mass turns into energy, then even a very small imbalance makes a very big boom. (The Hiroshima bomb blast was about the equivalent of 1 gram of matter converted into energy.) Whenever Professor McGonagall transforms herself from a human into a cat, she ought to release as much energy as all of the atom bomb tests ever done, all at the same time. And where does she get the extra mass when she turns back into a human?

There doesn't seem to be any good answer to this problem. If mass isn't conserved, maybe it is sloughed off somewhere during the transformation (yuck) or stored in some extra dimension or something. It is a vexation. You have even more problems if you are transforming material from one element to another: it can be done, but with difficulty. Consider Medusa's problem of turning people into stone. People are made up mostly of carbon and water, whereas stone is mostly silicon.

Carbon has atomic number 6 and atomic mass 12, and has six protons, six electrons, and six neutrons. Silicon has atomic number 14 and atomic mass 28, with 14 protons, 14 electrons, and 14 neutrons. You can't turn one elementary particle into another willy-nilly: to convert the carbon in a living body into silicon, you have to provide the extra particles somehow. Perhaps Medusa bombards her victims with high-energy particles? And if she can do that, why bother turning them into stone at all?

2.3 DISAPPARITION AND THE CONSERVATION OF MOMENTUM

> Morris got bugeyed. "You can teleport?"
> "Not from a speeding car," I said with reflexive fear. "That's death. I'd keep the velocity."
> —LARRY NIVEN, "THE FOURTH PROFESSION"

In the Harry Potter novels, the power of *disapparition* is used commonly. This is the ability to vanish from one spot and reappear instantly in another. In other works this maneuver is more commonly referred to as *teleportation*. In *Star Trek* the transporter is used for this purpose. Issues of conservation of energy plague teleportation in a similar manner to shape-changing. In disapparating, is Harry being converted to energy, zapped off somewhere at the speed of light, and converted back? If so, this is an awful lot of energy to manage: 9×10^{16} J for every kilogram we teleport. If even 1% of the energy isn't contained somehow, we have the equivalent energy of an H-bomb going off. We also have all the other problems from the last section: turning 80 kg or so of matter into "pure energy" violates all sorts of conservation laws, and also involves a rather bad issue related to entropy. Or murder, if you'd rather think of it that way. A human being is a very complicated structure, and by transforming the body into pure energy, as far as I can tell, you are killing the person. Bringing a person back to life after doing this by reconstituting him or her elsewhere is implausible.

Tearing someone apart *here* and putting that person back together *there* seems pretty hard. Another way in which authors have justified teleportation involves ideas of quantum mechanics. Larry Niven mentions this in his essay "The Theory and Practice of Teleportation" [178]. Quantum mechanical tunneling is a lot like teleportation: a particle, like an electron, goes from one side of a barrier to the other side without moving through the intervening space. Well and good. It works for electrons, why not for people?

To explore this idea, I am going to invoke the famous Heisenberg uncertainty principle. Most of the physics we've discussed so far has been classical: we've assumed that objects follow well-defined trajectories. In essence, we've assumed that we can know exactly where they are and how fast they move at all times. This isn't true. Very fundamental ideas of physics tell us that

$$(\Delta x)(\Delta p) \geq h/4\pi, \tag{2.1}$$

where Δx is the uncertainty in the position of the object, Δp is the uncertainty in the momentum $(= Mv)$, and h is Planck's constant, which has a metric value of 6.626×10^{-34} J-s (joule-seconds). This inequality means that we can't measure the position or speed of anything with arbitrary precision. Because we are taking the product, making the uncertainty in one smaller makes the uncertainty in the other larger. Unfortunately, h is really small. This is why quantum mechanical effects are important only for atoms or subatomic particles, at least under most circumstances. There are some fascinating exceptions, however.

For the past century physicists have performed experiments in which quantum mechanical effects have manifested themselves on macroscopic scales. Two older examples of large-scale quantum behavior are super-conductivity and superfluidity. At low temperatures, atoms in liquid helium behave in some ways as if they were a single atom, with no individual identity of their own. This superfluid state is characterized by almost no resistance to motion, helium liquid "climbing" out of containers, and weird quantum mechanical vortices occurring. Unfortunately, this superfluid state happens only when you can get the temperature below about 2 K above absolute zero.

Superconductivity is a similar low-temperature effect in which the resistance of metals to electrons flowing through them drops abruptly to zero at temperatures a few degrees above absolute zero. This happens more or less because the electrons in the metal find themselves in a superfluid state. In the early 1980s physicists found examples of substances that became superconductors at high temperatures. However, "high" is a relative term, meaning around the temperature of liquid nitrogen, 77 K. This is still about 200°C below room temperature.

In 1995, Carl Weiman and Eric Cornell at the University of Colorado at Boulder led a team that created the first Bose-Einstein condensate in atoms of rubidium. Wolfgang Ketterle's team at MIT achieved this a short time later. A Bose-Einstein condensate (BEC) is a group of atoms cooled to such a low temperature that their quantum mechanical uncertainty is so large that one cannot tell one individual atom apart from another one, even in principle. (Superfluid helium is like a BEC in some ways, but is much more complicated.) To get this state, the teams used a combination of lasers and other techniques to achieve temperatures of about 200 billionths of a degree above absolute zero!

In a BEC, the position of an atom is completely uncertain within a region a few hundred micro-meters across. This is roughly the diameter of a human hair. This doesn't sound big, but by comparison, the size of an individual atom is only about an angstrom, or 100,000 times smaller than that. It's a good start.

There is an effect called quantum teleportation in which the information about a quantum mechanical system can be transmitted from one quantum system to another. It has only been performed on systems of a few atoms at a time so far. As the webcomic *xkcd* points out, it isn't the same as "real" teleportation [172]. Unfortunately, there is another quantum mechanical effect called the "no-clone" theorem, which proves that it is impossible to create a perfect copy of a quantum mechanical system [255].

The common feature is that all of these effects occur at really low temperatures. In recent years physicists have been able to take large macroscopic objects and "cool" them to the point that quantum mechanical effects are important. (I put "cool" in quotation marks because the techniques that are used don't involve cryogenics or any of the classical methods used to cool large objects.) This is a long way from teleporting objects, but it is a start. The main examples are the mirrors used in LIGO, the detector designed to detect gravity waves from objects such as colliding black holes. The LIGO mirrors are "cooled" using lasers and electromechanical techniques so that their motion is limited only by quantum mechanical uncertainty. This is because they are being used to detect gravitational waves, which are so weak that gravity waves from two colliding black holes will make the mirrors move by a distance 100,000 times smaller than an atomic nucleus!

If we could only make h larger somehow, we might be able to build a practical teleporter. Unfortunately, two things stand in the way of doing this. First, h is a fundamental constant of nature. No one has any idea of how to change its value, let alone whether this is even possible. Second, even if we could change h, small changes in its value radically change the laws of chemistry. Changing its value by only 1% or 2% would probably make life impossible.

The idea of locally changing the value of Planck's constant has been used by Tim Powers in some of his urban fantasy novels, most notably in *Last Call* and *On Stranger Tides*, although not for teleportation [195][196]. In the latter book, an eerie scene takes place at the Fountain of Youth. Dr. Hurwood, the book's villain, states that the "uncertainty" is polarized there: the ground has none, while the air's quantum uncertainty is huge, to the point that a shadowy personality can answer questions from it. Powers is an author who likes to play around with pseudoscientific ideas like this in fantasy settings, which gives his works a uniquely creepy vibe.

Teleportation certainly seems like fantasy, in that I don't see any means of teleporting large objects short of magic. It violates too many laws of physics.

If we imagine that teleportation really works, however, what do the laws of physics imply about it? Larry Niven was the first writer to discuss the conservation of momentum as it applies to teleporting. The epigraph at the beginning of this section illustrates the point. Let's say you're in a moving car, and let's also say that you're a wizard in Harry Potter's London and have passed your disapparition test. You are being driven down the highway at 60 mph and see a friend on the side of the road, so you disapparate out of the car to her side. According to Einstein, all reference frames are equivalent, so in principle you keep the momentum you had from the car and appear by her side traveling at a speed of 60 mph relative to her and the ground. Ouch! In a similar vein, there is a serious issue in "beaming down" to a planet using a *Star Trek* transporter:

Let's say that the *Enterprise* or one of its descendants is in a geo-synchronous orbit around the Earth, so that it always stays over the same point on Earth's equator. You might think that this means that the *Enterprise* is moving at the same speed as the point Earth, but not so: the Earth makes one full revolution around its axis every 24 hours, so the spaceship must rotate around the Earth in the same amount of time. It moves through a larger circle in the same amount of time as a point on the surface of the Earth, so it must be moving faster. Because the geosynchronous radius is 42,000 km, it is moving faster by a factor of $42,000/6,400 = 6.6$, which is the ratio of the radius of the geosynchronous orbit to that of the Earth. Since a point on the Earth's equator moves at a speed of about 1,000 mph, anyone beaming down to the planet will land with a speed relative to the ground of $6,600 \text{ mph} - 1,000 \text{ mph} = 5,600 \text{ mph}$. This could be a problem.

Larry Niven has written some excellent stories that have been collected in the book *A Hole in Space*, which dealts with teleportation and its scientific and cultural ramifications [180]. In these stories and in the essay "The Theory and Practice of Teleportation," he discusses whether objects teleporting downhill increase their temperature, the conservation of momentum in teleporting to other latitudes, and similar ideas. We'll look at these ideas more in the web-based exercises, which can be found at press.princeton.edu/titles/10070.html.

On a side note, there's a lot of inconsistency in how *Star Trek* handles the transporter. Larry Niven in his collection *All the Myriad Ways* and Alfred Bester in his novel *The Stars My Destination* both noted how good a weapon a teleporter is [178][38]. The *Star Trek* transporter represents the epitome of this: using it, one can put a bomb anywhere. One can kidnap anyone. In one episode, a transporter malfunction regressed Picard to a ten-year-old child! Think of the ramifications: repeat this somehow and you have the Fountain of Youth. Maybe by changing the settings, you can heal any disease, any injury. And the show *never* explored any of the ramifications of this!

2.4 *REPARO* AND THE SECOND LAW OF THERMODYNAMICS

"Would you like my assistance clearing up?" asked Dumbledore politely.

"Please," said the other.

They stood back to back, the tall thin wizard and the short round one, and waved their wands in one identical sweeping motion.

The furniture flew back to its original places; ornaments reformed in midair, feathers zoomed into their cushions; torn books repaired themselves as they landed upon their shelves . . . rips, cracks and holes healed everywhere, and the walls wiped themselves clean.

—J. K. ROWLING, *HARRY POTTER AND THE HALF-BLOOD PRINCE*
[197 PP. 64–65]

There's a scene in one of Jim Butcher's Dresden Files novels in which Harry Dresden and his friends are being chased by a giant-sized scarecrow through the streets of Chicago [45]. It's just rained, and Harry needs to blast something in front of him, so he sucks energy out of a water puddle the giant is about to step in to use to fireball whatever is in front of him. The giant slips on the newly formed ice, the obstacle is destroyed and everyone, except the monster, and most physicists, is happy.

What Harry just did violated one of the fundamental principles of physics: the second law of thermodynamics. The second law of thermodynamics says:

In order to make heat flow from an object at a higher temperature to one at a lower temperature, you have to do work.

It certainly doesn't seem that Harry is doing any work in making the heat flow in the opposite direction it "wants" to go. Harry has violated the law that in any thermal interaction, the *entropy* of the world must increase.

2.4.1 Entropy Changes

Entropy is a subtle concept in physics. As most of my readers are aware, entropy is a measure of the "disorder" of the world. It is subtle because it is hard to formally define what order and disorder are. In lay terms, an ordered state of the world is one that is *less probable* than a disordered one. To give an example, let's say you are a college student. You have just cleaned your dorm room, so your books are all on the bookshelf, the covers are on the bed, blank paper is in the desk drawer, beers are in the fridge. This is an ordered state because it is an improbable one. Joking aside, we can imagine taking each item in the room—textbooks, beer bottles, bedcovers, and so forth—and tossing a die to see where to put everything. If the die lands showing 1, we put the item in the fridge; if it shows 2, we put the item on the bed; if it shows 3, we put the item in the bookshelf, and so on. When we do this, we are likely to find bedsheets on the bookshelf, beers on the bed, and books in the refrigerator. The ordered state is ordered because it is a low-probability one in that sense: tossing the die will far more likely end up with everything in the room thoroughly mixed up.

It is unfortunately a long and hard road to get from the concept of order and probability to heat and temperature. It can be done, but it is beyond the scope of this book. Suffice it to say, if we have a thermal system at absolute temperature T and add an amount of heat Q to the system, the entropy of this thermal system increases by an amount

$$\Delta S = \frac{Q}{T}. \tag{2.2}$$

If we remove the heat, the entropy decreases by the same amount. There's nothing that forbids the entropy of one part of a thermal system from decreasing, but the second law of thermodynamics tells us that it has to increase by at least as much somewhere else.

The reason why heat spontaneously flows from high temperatures to low temperatures has to do with the denominator in the formula. Because we are dividing by T, you get more entropy adding heat to a cold object than to a hot object. So a hot object in contact with a cold one will "want" to lose heat to the other one. Yes, its own entropy is lowered by losing heat, but the cold object's entropy increases by a larger amount. The total entropy in the world increases. If the process were to go in reverse, as in the Dresden Files novel, the net entropy would decrease. Let's try to put some numbers in.

   Harry extracts enough energy to freeze a large puddle. Let's assume that the water in the puddle has a mass of 100 kg. I'm going to simplify things by assuming that the puddle is already at 0°C (= 273 K), as befits a novel set in Chicago in midwinter. It is already at its freezing temperature, although still in a liquid state. All we have to do now is extract energy at constant temperature to freeze it. The latent heat of fusion of water is 3.34×10^5 J/kg. This is the amount of energy we must extract from each kilogram of water at its freezing point to turn it into ice. Therefore Dresden must extract 3.35×10^7 J from the puddle to freeze it. The entropy of the puddle has decreased by

$$\Delta S_1 = 3.34 \times 10^7 / 273 = 1.23 \times 10^5 \text{ J/K.}$$

This energy is put into the fireball. All Ray Bradbury fans know that paper burns at Farenheit 451, or 232°C, or 505 K. Its entropy has increased by

$$\Delta S_2 = 3.34 \times 10^7 / 505 = 6.6 \times 10^4 \text{ J/K.}$$

This is less than the entropy leaving the puddle, so, all other things being equal, the entropy of the world has decreased by the difference, or 5.66×10^4 J/K.

   Is there a way out? The astute reader has probably been wondering how refrigerators work if entropy must increase. The answer is that refrigerators must generate enough entropy by some means or another to overcome the deficit. Harry must be acting as the refrigeration unit here. Work must flow from Harry into the fireball to make up the deficit. By my calculations, Harry must expend a minimum of 2.8×10^7 J to make this happen. This is about the amount of energy he would consume in food over the course of three days.

   The *reparo* spell used by Dumbledore and Slughorn is no less problematic. It decreases perceptible disorder; it is easier to see that disorder is lessened than in the Dresden case, but harder to calculate the entropy decrease. We can do a very rough estimate of this. Let's say there are a certain number of places where we can put any of the things in the room—say, 1,000. We simply imagine dividing up the floor of the room into 1,000 different cells and tossing objects at random into each of them. (Don't like 1,000? Then divide it up finer if you want. It'll turn out not to make a huge difference).

Then let's say we have 100 objects to distribute around: a few books, crystals from the chandelier, pens, furniture, other items. We take one object and toss it into a random place. There are 1,000 different ways to do this. We take a second object and do the same: there are 1,000 ways to do this. For this calculation I am assuming that we can put more than one object into a given cell. Therefore, there are one million different ways of distributing two separate objects! If N is the number of objects we have and M is the number of ways we can distribute each object, the total number of ways of distributing all the objects, is Ω,

$$\Omega = M^N. \tag{2.3}$$

Therefore, there are $1{,}000^{100} = 10^{300}$ different ways to distribute all of the different objects at random in the room, and only *one* way to do it correctly. If we were able to toss everything out at random once per second, it would take on average 10^{300} seconds, or 10^{292} years, before everything was in its proper place. This is much, much longer than the universe has lasted and much, much longer than the stars will last.

If one digs into the issue, it turns out that the change in the entropy of the situation is given by the formula

$$S = k \ln \Omega = kN \ln M, \tag{2.4}$$

where $k = 1.38 \times 10^{-23}$ J/K. To do the calculation I am using the property of logarithms that $\ln M^N = N \ln M$:

$$S = kN \ln M = 9.5 \times 10^{-21} \text{ J/K}.$$

This is a small entropy change compared to what Harry Dresden had to achieve when fighting the monster. The reason why the change is low is because of the small number of objects involved: we only had to sort out 100 things into their proper places. Another way to put it is that it would take Dumbledore and Slughorn a few hours to sort out where everything went to straighten up the room. In Harry Dresden's encounter, he would need to "sort" hundreds of moles of particles, something that is much, much more difficult.

Rowling ignored a lot of ramifications of the *reparo* spell. Can it be used to heal injuries? Can it be used to make someone younger? After all, an increase in entropy is linked to the flow of time. We separate the future from the past by looking at the direction in which entropy increases. Time travel is possible in Rowling's world; is the *reparo* spell reversing entropy by reversing the flow of time somehow?

I'd like to end this chapter the way I started it, by looking at contrasting philosophies of how magic works in a given author's world. Superficially, the Dresden chronicles are a dark mirror of the Harry Potter books. There are a number of striking similarities between them, apart from being about wizards named Harry.

- They are set in the modern-day world with magic side-by-side with the mundane.
- The magic worlds are hidden from the "muggles" in each series.
- Both have draconian magic "governments," the White Council in the Dresden Files and the Ministry of Magic in the Potter books. Neither of the governments has much regard for the civil rights of its respective subjects.
- Both series center on conspiracies to undermine said governments.
- Both series use fake Latin for incantations.

However, Jim Butcher seems to have put more thought into making his system of magic self-consistent than J. K. Rowling has. As I discussed above, the *reparo* spell involves less of an entropy decrease than Dresden's spell did, and thus in a physics sense is less implausible. However, to me Dresden's spell *feels* less implausible than Dumbledore's. This is because Jim Butcher established rules for how magic is used throughout his books. In other works, Harry Dresden remarks that the rings and amulets he wears store up kinetic energy over time by robbing a little bit from when he walks. When they are discharged in a fight, they must be recharged. Harry Dresden gets exhausted from too much magic use, which never seems to happen in Rowling's novels. The plots in the Dresden Files novels are constrained and driven by the limitations on magic.

Rowling's books don't work the same way. Even though she writes in *Harry Potter and the Sorcerer's Stone* that Harry learns early on that not all of magic is wand-waving and saying a few words, this is pretty much how it works throughout her books. There don't seem to be any major rules on what is hard to do with magic versus what is easy to do. This is highlighted in the third book by the Ministry of Magic handing over a time machine to a thirteen-year-old girl, for no better reason than to let her take more classes. By contrast, in the Dresden Files novels, time travel into the past is a capital offense. Out of nowhere it is mentioned toward the end of *Harry Potter and the Deathly Hallows* that one cannot transfigure other items into food; this is probably driven by a plot need, to force the students in hiding in Hogwarts to have a connection to the outside world (i.e., the Hogshead tavern) [206]. I will reiterate that I enjoy the Harry Potter novels, but the purist in me gets annoyed by the lack of any overarching philosophy of what magic can and can't do. To paraphrase Poul Anderson, designing a magic system is the source of many plot points.

CHAPTER THREE

WHY HOGWARTS IS SO DARK

3.1 MAGIC VERSUS TECHNOLOGY

Many fantasy novels are set in preindustrial worlds, such as the Conan novels by Robert E. Howard, or in parallel worlds of quasi-medieval setting in which modern technology doesn't exist, such as the *Lord of the Rings* and *Earthsea* trilogies. Others are set in worlds where magic coexists with the mundane world. As an example of the latter, the Harry Potter books provide a means to examine the distinctions between the magical worlds and the mundane.

Hermione Granger in *The Goblet of Fire* mentions that most modern electronics won't work in Hogwarts because of the high concentration of magic there. This is a common feature in many works of the "urban fantasy" subgenre set in the modern world. Harry Dresden of the Dresden Files novels can't own a computer or a car with a modern electronic system because they routinely break when near him. Both Dresden's apartment and the houses of all wizarding families in the Harry Potter books are lit by candles. Most of the wizards in the Harry Potter books are improbably unfamiliar with the ways of modern-day muggles, to the point of not being able to operate a "fellytone." Given the lifestyle, why would anyone want to be a wizard? There are a lot of ways in which we could pursue this theme, but one way in particular has been overlooked by most writers: how dark a castle like Hogwarts or Harry Dresden's apartment would be if illuminated only by candles.

3.2   ILLUMINATION

Much of the action in the Harry Potter novels is set in a huge castle, Hogwarts, somewhere in Scotland. The castle is described in loving detail, with ghosts, moving staircases, and quasi-living pictures. The Great Hall at Hogwarts is one of the main settings in the castle. It is where the students eat, take important exams (such as their O.W.L.s), and learn disapparition. It is described thus when Harry Potter sees it for the first time in the novel *Harry Potter and the Sorcerer's Stone* [201]:

> Harry had never imagined such a strange and splendid place. It was lit by thousands and thousands of candles that were floating in midair over four long tables, where the rest of the students were sitting.

Even with the myriad candles, I suspect that the Great Hall would be a pretty dark place. One of the big but seldom remarked-on benefits of living in the modern world is the ability to light a dark room using electric lights. Herein hangs a tale.

Before Edison invented the electric lamp, all means of artificial illumination involved fire—burning something. If you've ever had to do your homework during a power outage you know that doing homework by candlelight is not the romantic, nostalgic affair it is painted as in historical novels. Much of the time you are maneuvering your book to catch as much light as possible from the flickering flames. No matter how bright the candle is, it is still pretty dim and very hard to read by.

This is a subtle issue that movies get wrong most of the time. Almost any historical movie will present a candlelit room as being nearly as bright as one lit by electric lamps. The reason for this is obvious: having characters stumbling around in what appears to be almost pitch blackness to modern viewers will detract from the story. However, the reason why candlelight and torchlight are dark is subtle.

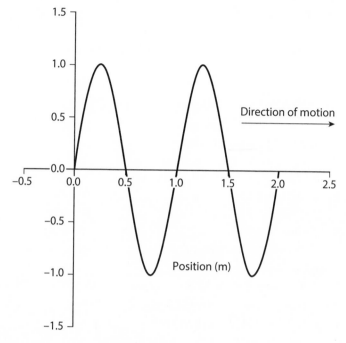

Figure 3.1. Wave with 1-meter wavelength.

To the human eye, not all light sources are created equal. Do the following experiment: laser pointers are available in both red and green varieties. Take two of equal power and shine both against a white screen or a light-colored wall. The green diode will seem much brighter than the red diode even though the power is the same. This is because the eye's rods and cones are most sensitive near the wavelength of green light and less sensitive to the red. These retinal cells can detect light over only a very small spectral range; everything outside that range is invisible to the human eye.

To discuss this, we need to talk about how light works. Light is an electromagnetic wave: think ripples on a pond, not ocean breakers. I've made a picture of how a wave "looks" in figure 3.1.

The wave is moving to the right: it is a ripple of something, that something depending on what the wave is. If it's a water wave, it's a little hill on the water. If it's a sound wave, it's a compression of the air. If it's a light wave, it's a ripple in the *electromagnetic field*, which is a bit more complicated to explain. Here goes:

- Normal matter is composed of three different types of particles: protons, electrons, and neutrons.
- Electrons and protons have charge; by convention, the proton charge is positive and the electron charge is negative. The neutron, as the name implies, has no charge (it is neutral).
- Like charges repel, unlike charges attract.

So far, this is what most of you learned in basic chemistry class. However, what they didn't teach you was this:

- Take a charge over *here* and wiggle it up and down somehow. A charge over *there*, some distance away, will begin to wiggle up and down in sympathy.
- However, it takes a while for the second charge to begin to wiggle up and down. If the distance is 1 meter, the second charge begins to wiggle up and down $1/299{,}792{,}458$ of a second later.
- The time it takes is greater for bigger distances, lesser for smaller ones: the relation between the time it takes, t, and the distance, x, is given by

$$x = ct, \tag{3.1}$$

where c is the speed of light. By definition,

$$c = 299{,}792{,}458 \, \text{m/s}.$$

As far as anyone knows, this is the fastest speed anything can travel at in our universe.[1]

We like to say that there is a something that travels between the two charges. We call this something a *light wave*. What we measure is simple: charge wiggles up and down over here, charge over there wiggles in response after a time delay. The charge we wiggle here loses energy, the charge over there gains energy. Because we like to think energy is conserved, we say the light wave carries the energy from the first charge to the second charge. This is what light is, in a fundamental way: when you look at a star, charges in the star are wiggling up and down. Many years later, charges in the retina in your eye wiggle up and down in response, causing electrical signals to be sent to the visual centers of your brain, producing the sensation of sight.

However, the eye is not sensitive to all light. To understand this, we need to introduce two key concepts:

- The frequency of the light wave, f, is the number of times per second we wiggle the charge up and down. It is measured in cycles per second, or Hertz (Hz).
- The wavelength, λ, of the light wave is the distance between adjacent crests of the wave (see fig. 3.1). The letter λ is the Greek lowercase lambda, corresponding to lower case "l" in the Roman alphabet.

Wavelength and frequency are related in a simple way:

$$f \times \lambda = c. \tag{3.2}$$

Thus, a light wave with a frequency of one cycle per second will have a wavelength of 300 million meters. This is a very long wavelength, much longer than the eye can see. In principle, and in practice as well, light can have any wavelength.

Most sources of light contain a mixture of different frequencies. (Lasers are an exception to this rule.) How bright a given light source appears depends not on the total power that the source puts out but on how much it puts out in the spectral region in which the eye can see. An infrared laser may have a total power of several kilowatts, but since the light isn't visible to the eye, it wouldn't work at all as a source of illumination. The eye is sensitive to light over a very small spectral region, wavelengths very roughly from 350 nm to 700 nm. A nanometer (nm) is a very small distance—1 nm equals one-billionth of a meter, or roughly 10^{-5} the diameter of a human hair! Short wavelength means high frequency: a wavelength of 600 nm means a frequency given by

$$f = c/\lambda = 3 \times 10^8 \, \text{m/s} / 600 \times 10^{-9} \, \text{m} = 5 \times 10^{14} \, \text{Hz}.$$

Light of this wavelength will appear orange to the eye. In general, light of wavelengths 350–500 nm will appear purple or blue, light of wavelengths 500–600 nm will span the gamut from green to yellow-orange, and light of wavelengths 600–700 nm will appear orange or red. In this spectral region the eye is most sensitive to light of wavelength 555 nm, which is green light.

Sources of light such as candles or torches or light bulbs rely on heat to create light; in physics terms, they are *blackbody radiation sources*. Blackbody radiation sources are very important in this book; for example, the stars are all essentially blackbody radiators. A thermal source can be characterized by two factors, its radiating area and its temperature. The Stefan-Boltzmann formula tells one how much power (energy per second) a thermal source emits:

$$P = \sigma A T^4. \tag{3.3}$$

The terms appearing in the equation are:

- P: the emitted power (units of J/s, or watts);
- A: the emitting area (units of square meters, m^2);
- T: absolute temperature (i.e., measured from absolute zero) (units of Kelvin, K); and
- σ: the Stefan-Boltzmann constant: $\sigma = 5.67 \times 10^{-8}$ W/m^{2}K^4.

This light is distributed over a wide range of wavelengths. Figure 3.2 shows the blackbody spectrum due to several different sources of light. In particular, it can be seen that the curves are small at both short and long wavelengths but go through a peak in the middle. The peak is determined only by the temperature:

$$\lambda_{peak} = 2,900 \,\text{nm} \times \frac{1,000 \,\text{K}}{T}. \tag{3.4}$$

This isn't the fundamental form of the equation but a version that makes it easy to calculate peak wavelengths. For example, the sun's temperature is 5,800 K, implying a peak wavelength given by

$$\lambda_{peak} = 2,900 \,\text{nm} \times \frac{1,000 \,\text{K}}{5,800 \,\text{K}} = 500 \,\text{nm}.$$

This says that the higher the temperature of the light source, the shorter the peak wavelength will be. In addition to blackbody spectra, figure 3.2 also shows the relative sensitivity of the eye to light of different wavelengths; note that the wavelength where the eye is most sensitive (550 nm) is very close to where the peak of the Sun's spectrum is (500 nm). The other light sources are much cooler than the Sun, meaning that the overlap isn't as good. This is the point of this section.

Engineers use the unit of a *lumen* (lm) to measure how bright a light source is: because the eye is not uniformly sensitive to all wavelengths, there is no direct conversion of lumens to watts. The conversion and the *luminous efficacy* of the light source depend on the properties of the given source. There is an online appendix for this book (press.princeton.edu/titles/10070.html) describing how to calculate the brightness of a given light source from the blackbody spectrum. Readers who are interested in it can read it, but in general, the better the overlap of the spectrum with the eye's sensitivity, the brighter the light source will be. The *luminous efficacy* characterizes how bright a given source of light will be:

$$B = LP, \tag{3.5}$$

where

- B is the brightness of the light source (in lumens);
- L is the luminous efficacy (in units of per watt, lm/W); and
- P is the total power radiated by the light source (in units of watts).

The highest possible value for L is 683 lm/W; you can only get a value this high if you are using a laser that emits light with a wavelength of 555 nm. The efficacy of any blackbody source, even the Sun, will be much lower. The key parameter is the temperature T of the different sources. My friend Dr. Raymond Lee is an expert on natural and artificial illumination. At my request he measured the spectral emissivity and blackbody temperature of two artificial light sources, a 75 W tungsten bulb and a candle. From his measurements, we determined that the bulb had an effective blackbody temperature of 2,740 K and the candle a temperature of 1,850 K. It was also possible to determine the total power radiated by the candle in all spectral regions. Using the data, we calculated a total radiated power of 15 W, with an accuracy of about ±15%.

Table 3.1 shows the luminous efficacy of various different light sources.

Table 3.1
Luminous Efficacy of Various Light Sources

Light source	T (K)	Luminous Efficacy (lm/W)	Power (W)	Brightness (lm)
Sun	5,780	93	N/A	N/A
Tungsten bulb	2,740	13	75	975
Candle	1,850	0.8	15	12

The candle will be much dimmer than the bulb for two reasons:

1. It radiates away only one-fourth the power the light bulb does.
2. Much less of that power is in the visible region of the spectrum.

The second point is the more important one. Because the blackbody temperature of the candle is so low, most of its light will be emitted in the infrared region, where it is invisible and useless for illumination. Only about 1% of the light from the candle is radiated away in the visible region of the spectrum, compared to about 10% of the light from the bulb. In figure 3.2 I've graphed how well the spectrum from each light source overlaps the eye's sensitivity. The overlap with the solar spectrum is extremely high. It is less so for the tungsten bulb, and almost nil for the candle. Thus, a candle will appear 80 times dimmer than a light bulb! [2]

Let's return to the question that inspired this section, how brightly illuminated is the Great Hall? How well a light source can brighten an area depends on the *illuminance* of the light, which is the total luminance divided by the total area it is illuminating. That is to say, a huge stadium with a single candle suspended above the middle of it will be much dimmer than a broom closet illuminated with an arc lamp. So we need the total area of the Hogwarts Great Hall. As I write this, I am sitting in the main reading room of the Library of Congress; it seems about the size of the Great Hall, although a different shape. Pacing it off, it is about 30 m in diameter (i.e., about 100 feet), for an area of about 700 m². The illuminance of the hall is the total luminant flux divided by the area it is illuminating. To get an exact value would be very tricky, but we can

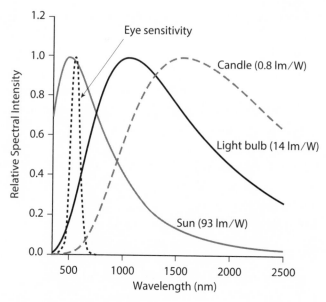

Figure 3.2. Spectra of various light sources.

reason as follows. Assuming that the candle is a uniform illuminator, equal amounts of light will be going off in all directions. Half of it will be going upward and will be wasted for purposes of illumination. Of the rest, some part will be going off to the side, although some of that will be reflected off the walls and not entirely lost. I'm going to estimate that 40% of the light will be useful for seeing with. Therefore, the total illumination (I) is given by

$$I = \frac{0.4 \times 5{,}000 \text{ candles} \times 12 \text{ lm/candle}}{700 \text{ m}^2} = 34 \text{ lm/m}^2.$$

This is an average value: some areas will be somewhat brighter, others somewhat dimmer. However, it's not going to vary by an enormous amount. This illumination is inadequate for reading and writing, which usually require a minimum value of 200 lm/m^2. The Great Hall is used for eating, but even for simple repetitive tasks a minimum of about 100 lm/m^2 is required.

Of course, the Great Hall is bewitched, so that the ceiling shows the sky outside; during the day it'll be pretty bright. Maybe the reason why the four founders bewitched it in the first place is the dimness of artificial illumination in medieval times. Solar illumination can reach values in excess of $10^5 \, \mathrm{lm/m^2}$, or over a thousand times brighter than the candlelight available in the Great Hall. However, at night the candles will provide most of the light, as the nighttime sky in the country can get very dark. Even the full Moon provides an illuminance of only $0.2 \, \mathrm{lm/m^2}$. Please note that this is not a measure of how bright the full Moon will appear when, say, Ron *looks at it*; this is a measure of how much lighting the full Moon can provide for tasks such as eating and reading. More or less, the illumination of the Great Hall at night by the candles will be about the same as that experienced by someone standing outside during the middle of what we in the light-and-color biz refer to as civil twilight: the period of time right around when the sun is setting below the horizon.

Thinking about this issue, one starts to wonder about a few things more. First, candles have to be held upright to burn correctly, but there is a shadow region directly under the flame (where the candle is). This is going to block even more light from getting to the tables.

Second, the cost of using candles in this fashion is going to be horrific: generally speaking, they're probably going to have to replace each candle each day. If we assume a cost of $1 per candle, this represents a bill of $5,000 per day, or (assuming a 200-day school year) a $1 million annual lighting bill for the Great Hall alone. Conservatively, it'll cost about ten times more for the entire castle, or about $10 million per year. In magical terms, this is about one million golden galleons per year, or perhaps 1,000 galleons per student per year. By comparison, lighting a building using electricity is much cheaper. Under the same assumptions as above, we can provide an illumination of $200 \, \mathrm{lm/m^2}$ to the Great Hall using a total power of about 10 kW using tungsten bulbs. This is enough for eating, writing, or reading. Assuming electricity costs of $0.1/kW-hr, lighting the hall costs about $10 per day for ten hours of illumination, or fully 500 times less than using candles. (How much do those students pay to go school?) Finally, as these candles burn, hot wax should drip onto the students eating below. I'm a little surprised Draco never complained about this.

One response to this entire chapter is to say, "Well, they're *magic* candles, so they burn brighter than ordinary ones, and don't drip, and cost nothing." This is perfectly acceptable, although J. K. Rowling never writes that they are. It doesn't detract anything from the story to take this view. However, another response is to say that this is one detail of a fascinating book, and it is part of "the Game" to examine such details and see where they lead us. I feel that the latter approach is far more interesting and more fun than leaving the books unexamined.

NOTES

1. We will usually round this to $c = 3 \times 10^8$ m/s, which is almost always close enough for any calculation we make.

2. Of course, tungsten bulbs themselves aren't ideal illuminators, which is why people are switching to compact fluorescent bulbs. The fluorescents have higher effective temperatures, so even more of their light is in the visible spectrum.

CHAPTER FOUR

FANTASTIC BEASTS AND HOW TO DISPROVE THEM

4.1   HIC SUNT DRACONES

One of the chief hallmarks of a fantasy story is the presence of other-worldly creatures. This is true of science fiction as well. In science fiction they are called aliens rather than monsters, but in numerous works the two are treated in a very similar manner. The troll at the bridge judges whether the party is worthy to cross or whether he must eat them; the aliens from Alpha Centauri judge humanity and decide whether to destroy us or reward us with advanced technology.

The chief difference, at least between "hard" science fiction—that based on known scientific principles—and fantasy, is that aliens in science fiction need to "work" somehow; that is, their biology must function along known and understood scientific lines. This does not mean they must be dissimilar to monsters or other creatures in a fantasy period. For example, in Robert Heinlein's novel *Starman Jones*, the aliens attacking the human settlers on an unknown world look very much like centaurs, hybrids between humans and horses [122]. Another Heinlein novel, *Glory Road*, features beasts that are very much like dragons (i.e., large firebreathing lizards) except they cannot fly [116]. Of course, this novel has very strong fantastic elements. There are countless examples from other stories in which alien creatures are based on mythological or fantastic animals.

This raises an interesting question: which fantasy animals are possible ones and which aren't? That is, if we took a census of fantastic beasts from various fantasy novels, which could we then place in a science fiction story without having to bend the rules too much? Here's a very small list of fantastic creatures from the Harry Potter novels (not, of course, unique to them): centaurs, giants, unicorns, giant spiders (*acromantula*), hippogriffs, and dragons. Which of these are possible beasts and which are impossible? To answer the question we need to look at the issue of scaling and how it affects the properties of animals and their metabolism.

4.2 HOW TO BUILD A GIANT

The recent movie *Avatar* is set on the world of Pandora, a moon of a gas giant circling one of the stars in the Alpha Centauri system. The surface gravity on Pandora is about 80% Earth normal; the Pandorans are portrayed as large, willowy beings standing about eight feet tall and possessed of great physical strength compared to humans. Is this accurate? Given the facts above, how large should a Pandoran be?

There is a lot to this question, and many aspects of biology enter into the answer. However, the field of biomechanics can guide us in a general consideration of this problem. If we want to build a giant, how do we go about it? Can we simply take a human and, say, double him in all dimensions? Or is it more complicated than that?

In the early 1920s, J. B. S. Haldane, the great mathematical biologist, wrote the fascinating essay "On Being the Right Size" [106]. In one part of it he examined whether giants such as Giant Pagan from *The Pilgrim's Progress* could stand. This is an interesting point: if we look at organisms from the small (say, insects) to the largest land animals, such as elephants, we realize their shapes are not particularly similar but do change in a systematic way. The legs supporting insects are very narrow and spindly compared to their body size, while the legs supporting an elephant or other giant creature are very thick compared to its body. Also, there are no creatures that stand on two legs, even for short periods of time, that are larger than humans. This is an aspect of a famous law originally formulated by Galileo and usually referred to as the square-cube law: *the weight an animal must support is proportional to the cube of the size of the animal, but the supporting structure is proportional only to the square of the size.* There are a lot of ramifications to this law, and it is not the only scaling law that applies to biology, so we will go through it slowly and in detail.

Let's model an upright biped (human, Pandoran, or giant) as a large load standing on two pillars. This will do for a static model; what happens when the thing walks I'll defer until later. How much weight can be put on the pillars before they collapse? As we will see, this determines the overall structure of the biped.

Let's model a human as a cylinder (torso and head) standing on two cylindrical pillars (the legs). We're ignoring the arms and head, but I'm assuming that their weight is not significant compared to the weight of the torso. I'm also going to ignore the weight of the legs. It's an oversimplified model but should be good to start us out. I'm going to assume that the length of the torso (L) is the same as the length of the legs, and that the radius of the cylinder is about one-sixth the length, meaning that the circumference of the cylinder is very roughly equal to its length; this seems reasonable, given average belt lengths. This is a very rough sketch of a human being, obviously.

The volume of the torso is given by

$$V = \pi r^2 L = \pi L^3/36 \approx L^3/12, \tag{4.1}$$

and the mass is $M = \rho V$, where ρ is the average density of the human body, very roughly equal to the density of water ($1{,}000 \, \text{kg/m}^3$). This seems a reasonable assumption for the overall shape of a human torso: a simple calculation of the mass for a human whose overall height $2L = 1.8 \, \text{m}$ gives a torso mass of 60 kg, or 130 lb, which seems about right.

What we are interested in here is the width of the support pillars, the legs. The big question is one of scaling: as we increase the mass from that a normal man to that of a giant, will the overall shape stay the same? That is, will the width of the legs *with respect to their length* stay the same, or should we make them thicker, as Haldane implied? And can we estimate what the maximum height for such a giant could be, both on Earth and on other planets (such as Pandora), with different values of g?

There is one other question one can ask: why did we make the assumptions that we did, that is, that the overall leg length is equal to torso length, and that the torso diameter is proportional to torso length? We'll address that below as well.

4.2.1 Biological Scaling

The principle of *allometric scaling* is an extremely important one in biomechanics. Allometry is any relation that can be written between two biological variables in the form of a power law:

$$y = ax^b.$$
(4.2)

Quite often the independent variable is the mass, mainly because it is extremely easy to measure. For example, let's consider the relation between the mass and the length of the "cylinder torso" we modeled above. We can model the dependence of the length of the cylinder on its mass as

$$L \text{ (m)} = 0.22 M^{1/3} \text{ (kg)}.$$
(4.3)

We expect a relation of this sort between body length and mass, as mass is proportional to volume and volume has dimensions of $(\text{length})^3$. Another way to put this is that if we take two objects of the same shape but different overall scale, the ratio of their volumes is the cube of the scaling factor: as the size doubles, the volume and mass increase by a factor of 2^3, or 8.

This relation holds exactly only for objects that are the same shape. Animals, of course, come in all shapes and sizes. However, studies relating animal mass to overall length show that over a very wide range of sizes and species, the law that overall length is proportional to the cube root of mass holds relatively well. The surface area is related to the mass by the scaling parameter

$$A \propto M^{0.67},$$

again, consistent with geometrical scaling.

Back to our giant. A column such as a leg bone will buckle if the weight it supports exceeds a critical value. The great and prolific mathematician Leonhard Euler was the first person to investigate this problem, which is consequently referred to as Euler buckling: a column with a circular cross section of radius r and length L will buckle if the weight atop it exceeds a critical value, given by

$$W_c = \frac{\pi^3 E r^4}{4L^2}. \tag{4.4}$$

Here, E is the elastic modulus of the material, which is a measure of how "stretchy" or "bendy" the material is. I am going to assume that the leg proportions are set by this relationship: animal support-bone lengths and widths will be set by the criterion that they need to be able to support the weight on top of them safely and without buckling. Instead of solving the problem exactly, I'll discuss some scaling issues here. Since the weight is proportional to the volume, which scales as L^3, we can write out a proportionality,

$$L^3 \propto \frac{r^4}{L^2}. \tag{4.5}$$

Solving this,

$$r \propto L^{5/4}.$$

Another way to put this is that the ratio of radius to length, r/L, is proportional to the fourth root of the length $L^{1/4}$. As the being increases in size, the width of the legs will grow out of proportion to the rest of the structure. Giants should appear squat compared to normal human beings because they need thicker supports in relation to their height to hold up their weight. One can see that this is true by analogy with quadrupeds: elephants don't look like scaled-up horses; their legs seem a lot thicker relative to body size than a horse's legs. Grawp the giant, first introduced in *Harry Potter and the Order of the Phoenix*, is described as being 16 feet tall, or about three times the height of a normal human, if we round up [204]. His legs should therefore appear $3^{1/4} = 1.3$ times thicker than a human's in relation to their length. To be blunt, this calculation surprised me, as I would have expected the legs to be much thicker. Of course, this model is simple-minded in that I have assumed that the length of the leg is simply proportional to the overall body length, which is questionable.

There's an interesting supporting piece of evidence for this model. Bone mass is proportional to bone volume, which scales as $r^2 L \propto L^{14/4} \propto M^{1.17}$. That is, the bone mass (at least for leg bones) should increase more rapidly than the overall mass of the creature. This is in fact measured: bone mass increases as $M^{1.08}$. This is less than our estimate indicates, but perhaps not too far out in left field. This may simply indicate a failure of geometrical scaling for our overly simplified model of a biped.

One often sees the assertion that giant humanoids (at least on Earth) are impossible because of square-cube arguments, that is, because the weight increases so much more rapidly than the surface area supporting it. A recent paper in *Physics Education* states (in discussing the maximum height reached by an animal):

What would be the minimum required thickness of the bones of animals? To answer this question, let us examine the resistance of the bones and the weight-bearing limbs. Compressed solid materials may undergo a maximum tension, T_{br}, before breaking. [70].

The authors then go on to state that because of this rule, bone diameter should scale as bone length to the 3/2 power. This is simply untrue: the scaling law quoted above for bone mass implies that bone diameter scales at a lower power ($5/4 = 1.2$) with bone length, and again, studies of bone mass in relation to body mass bear out a lower scaling power. However, there may be a nugget of truth in the statement. The model we have put together assumes that the scaling between bone length and bone diameter is caused by the need to avoid Euler buckling of the structure. At low masses, the force needed to cause the structure to buckle is less than the force needed to break the bones in compression, but at some critical value of the height/mass ratio, this criterion is reversed. This is worked out in detail in one of the problems on the book's website (press.princeton.edu/titles/10070.html).

We can do a few other interesting calculations. For example, the walking speed of a biped is set by the pendular frequency of the swing of its legs. Detailed arguments can be found in Steven Vogel's textbook, *Comparative Biomechanics*, but essentially, the period of a pendulum is proportional to $\sqrt{L/g}$, where L is the length of the pendulum and g is the acceleration of gravity [243]. Viewing the moving leg as a swinging pendulum of length L, the stride length will be proportional to L, so the speed will be proportional to $L/\sqrt{L/g} \propto \sqrt{gL}$. If we assume that a typical human walking speed is about 4 mph (about 1.8 m/s) and that L is about 1 m, we arrive at the formula

$$v_w = 0.6\sqrt{gL}. \tag{4.6}$$

A few things to note:

- Larger animals have greater walking speeds than smaller ones. Interestingly, the time it takes to make a stride increases with size, but this is more than compensated for by the increased length of each stride.
- Higher-gravity planets will have faster walkers because of the factor g in the equation.

Grawp, being three times the size of a human, should be able to move 70% faster than one, although the movements will seem labored because of the time it takes for each stride.

Now, back to the Pandorans: they are giants compared to humans. The males' average height is 3 m, or roughly 10 feet. By comparison, a typical human height of 6 feet is about 1.9 m. Their planet has only 80% of the surface gravity of Earth. Because of this, human walking speed on Pandora should be about 10% less than on Earth because of the lower gravity. This should be noticeable, but the movie doesn't show people walking more slowly. However, Pandorans on Pandora should move 30% faster than humans because of the size difference. One should note that these results are dictated by simple physics and by the fact that Pandorans seem to walk the same way humans do.

Are these scaling laws correct? One must be a bit careful when applying them. The results should be taken with a grain of salt, for several reasons:

- First, these considerations are dictated by general physical principles, and may not be valid under local, specific conditions.
- Second, technically speaking, these scaling laws should apply only when comparing different species with similar shapes. That is, we might expect these laws to apply when comparing humans to (hypothetical) giants (i.e., two different species of bipeds), but not when comparing humans to horses or horses to eagles. However, no two species are ever exactly scaled versions of one another, so the results above have to be taken with a grain of salt.
- Finally, all of these scaling laws have to be assessed against actual experimental data, which can be tricky. It is hard to measure these scaling parameters with precision; usually they are quoted with error bars of about 10% or more.

Actual scaling parameters from studies done on real animals indicate that the arguments given above are within the bounds of possibility, but not proven. There is simply too much smear in the actual data. In particular, body surface area scales with mass to a power somewhere between 0.63 and 0.67. The higher exponent is what one would expect if geometric scaling were the only issue involved, and we did not need to worry about the support forces on the structure.

4.3 KLEIBER'S LAW, PART 1: MERMAIDS

> "The whale is not a fish, you know—it's an INSECT."
> —MONTY PYTHON: *THE SECRET POLICEMAN'S BALL*

4.3.1 Mermaids Aren't Fish!

I have made perhaps one of the most important discoveries in the field of cryptozoology in the last century: *mermaids aren't fish*. Perhaps I should say "merpeople" instead.... This actually seems a fairly obvious conclusion, at least when one looks at almost all depictions of mermaids in popular culture. The big clue is that merpeople, in almost any depiction you see them in, have horizontal flukes. This means they are actually oceangoing mammals, like whales, and not fish at all. Perhaps I should explain.

Herman Melville is best known for his massive fantasy novel, *Moby-Dick*. There are no mermaids in the story, but *boy* are there plenty of descriptions of whales. In the chapter "Cetology" he enters into a long discussion of whether the whale is a fish:

> To be short, then, a whale is A SPOUTING FISH WITH A HORIZONTAL TAIL. There you have him.... A walrus spouts much like a whale, but the walrus is not a fish because he is amphibious. But the last term of the definition is still more cogent, as coupled with the first. Almost any one must have noticed that all the fish familiar to landsmen have not a flat, but a vertical, or up-and-down tail. Whereas, among spouting fish the tail, though it may be similarly shaped, invariably assumes a horizontal position.

Melville places whales among the fishes, but he cites a number of reasons why biologists don't: the whale has lungs, not gills; the whale lactates and gives birth to its young, rather than laying eggs; and the whale spouts. And, as in almost all depictions of mermaids, it has a horizontal tail (or fluke), whereas all true fish have vertical tails. This implies a vastly different method of swimming for them: sharks, for example, swim by wriggling their fins and bodies back and forth, in a motion almost impossible to emulate by a swimming person. A whale, on the other hand, moves its tail up and down, which can easily be done by any swimming human. This is because the whale is a seagoing mammal. Millions of years ago the ancestors of the whale left the land and went back into the oceans. Their hind legs fused together to form the tail, which kept its skeletal structure more or less intact, meaning that the joints in it are hinged for up-and-down motion. Indeed, the cartoonist Chuck Jones, when trying to teach animators how to animate seals swimming, took one of his grandsons and tied his forearms to his torso and his two legs together, then put flippers on hands and feet and tossed him into the pool [131]. The natural motions the boy made in keeping afloat were very similar to how seals and all other sea mammals swim.

It seems clear from these arguments that merpeople must be a species of hominids that left the land thousands or perhaps millions of years ago and adapted to the waters. This seems even clearer when we consider other...ahem..."interactions" that sailors have claimed to have had with these creatures. The best story detailing a liaison between a human and a mermaid is "The Professor and the Mermaid" by Giuseppe di Lampedusa, author of the novel *The Leopard*. In the story, the eponymous professor meets a mermaid while a graduate student studying the classics; the mermaid Lighea is described as a combination of goddess and animal, and the aging professor is implied to have returned to her by jumping off a ship at the end of the story. Other similar stories exist, the best known being "The Little Mermaid" by Hans Christian Andersen, in which a mermaid falls in love with a handsome sailor. This doesn't seem too plausible if she were some species of fish. In the Disney version of the story, Ariel is just a girl with a tail, though a tail with clearly horizontal flukes.

Interestingly enough, the mermaids in the Harry Potter stories appear to be fish. In the book *Harry Potter and the Goblet of Fire*, they are described as having gills; in the movie, although it is a little hard to tell, careful examination shows that they have *vertical* tails that move back and forth rather than up and down [203]. It makes one suspect there are two different unrelated species out there that, by chance or convergent evolution, look remarkably similar. The picture of the mermaid in the Prefect's bathroom in the same film shows a much more hominid-looking mermaid, complete with horizontal flukes.

4.3.2 Kleiber's Law, Metabolic Rate, and Lung Capacity

If merpeople are really seagoing mammals, how long can they stay underwater? All seagoing mammals, such as whales and dolphins, must come up for air from time to time. Let's try to calculate this by making two assumptions:

- The metabolic rate of any animal is proportional to the rate at which the animal can breathe in air (technically, the mass or [equivalently] volume flow rate.
- The metabolic rate is proportional to $M^{3/4}$.

The first is easy to understand: metabolic processes need oxygen to proceed. The metabolic rate, or the rate at which energy is burned by the body, must be proportional to the amount of air taken in, divided by the time between breaths. The second assumption is known in biology as Kleiber's law, after the biologist who first formulated the rule. There is a good deal of evidence to support this rule over a very large range of body masses, although there is some debate over the exact scaling exponent [243, pp. 62–63]. To the best of my knowledge the fundamental reason for Kleiber's law is unknown.

The metabolic rate is proportional to lung volume multiplied by the average rate at which the animal breathes:

$$\mathcal{M} \propto V_L \times R_B, \tag{4.7}$$

where $\mathcal{M}$ is the average metabolic rate, V_L is lung volume, and R is the average number of breaths the animal takes per second. One tricky part: when whales breach the surface, after they blow water out their blowholes, they typically take a large number of breaths before diving again. This allows them to stay underwater longer.

Studies show, unsurprisingly, that lung volume is proportional to body volume, that is, mass; assuming this, and Kleiber's law, we find that, according to Walter Stahl [224],

$$R_B \propto M^{0.75}/M^1 \propto M^{-0.25}.$$

As the animal gets larger, the number of breaths per second it needs to take gets smaller, which explains why whales can stay underwater for long periods of time. Experimental data support this relationship. Stahl's paper gives the formula [224]:

$$R_B = 53.5M^{-0.26} \text{ breaths/min} \qquad (4.8)$$

for the average respiration rate as a function of mass (in kg). This formula gives a respiration rate of 17 per minute for an 80 kg adult male, which seems right: one breath every three seconds or so. Because of the water's buoyancy, merpeople could be significantly more massive than people. If we assume a mass five times that of a human, or 300 kg, the respiration rate works out to be 12 per minute. That is, respiration rate scales only slowly with body mass. We therefore expect merpeople to stay under the surface of the water only about 40% longer than the average human (if my assumptions about mass are correct). Adaptations to the water may enable them to stay underwater for longer times, just as long as they hyperventilate (for want of a better word) when they surface. However, there seems to be no way they can stay underwater for the extended periods of time indicated in movies and fairy tales. An analogy: U-boats were German submarines of World War II. They were small craft with diesel engines and could remain submerged for only a few hours at a time. Today's nuclear submarines are huge craft and can stay submerged for weeks at a time. Merpeople are to whales as U-boats are to today's nuclear submarines.

4.4 KLEIBER'S LAW, PART 2: OWLS, DRAGONS, HIPPOGRIFFS, AND OTHER FLYING BEASTS

> What is your name? What is your quest? What is the airspeed velocity of an unladen swallow?
> —OLD MAN, *MONTY PYTHON AND THE HOLY GRAIL*

To quote J. B. S. Haldane:

It is an elementary principle of aeronautics that the minimum speed needed to keep an aeroplane of a given shape in the air varies as the square root of

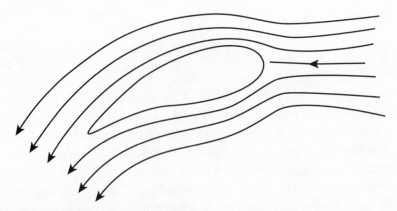

Figure 4.1. Airflow over a wing surface.

its length. If its linear dimensions are increased four times, it must fly twice as fast. Now the power needed for the minimum speed increases more rapidly than the weight of the machine. [106, p.18]

To explore this point in detail, let's consider air flowing over a wing with area A at speed v; this can represent any object flying at speed v through still air. Because of the shape of the wing, a lift force is generated upward on the wing (fig. 4.1). The lift is due to the fact that air is deflected downward by the wing, implying a force that pushes the air downward. By Newton's third law, there must be an equal but oppositely directed force pushing up on the wing. The expression for this force is

$$F_L = \frac{1}{2} C_L \rho A v^2, \tag{4.9}$$

where ρ is the density of the fluid the wing is moving through (1 km/m^3 for air), A is the area of the wing surface, and C_L is a dimensionless lift coefficient that is usually about equal to 1. It varies with the shape of the wing and the speed, but not enormously. We can take it as 1 for our purposes here. In steady flight, neither rising nor descending, the lift must be equal to the weight. We can equate the two expressions and solve for the speed required:

$$v = \sqrt{\frac{2Mg}{C_L \rho A}}, \tag{4.10}$$

where M is the mass of the creature or plane and g is the acceleration of gravity. The speed therefore depends on the ratio of the mass to the surface area of the creature. If L represents some measure of the length of the beast, then $M \propto L^3$, but $A \propto L^2$, so $v \propto \sqrt{L^3/L^2} \propto \sqrt{L}$. Bigger beasts must fly faster, proportional to the square root of body length. A Roc, a bird larger than an elephant, would have a flight speed of more than 100 mph just to keep it aloft.

These formulas can be used for science fiction stories as well. For example, in *Childhood's End*, the Overlords are a race of winged humanoids who live on an artificially low-gravity world [52]. They are somewhat larger than humans, about 7 feet tall, and presumably their mass follows the same scaling laws. A human visiting their world wasn't too discomfited by the their world's acceleration of gravity. Let's assume that it's about 80% of the acceleration of gravity on Earth. I estimate that their flight speed would need to be about 35 m/s (76 mph) to stay in the air, but could they fly at all? This is the question for the next section.

4.4.1 Metabolic Rates and Flying

The biggest problem with flight in large animals is that the metabolic rate for a large animal cannot be great enough to sustain the power needs for flying. This is indirectly related to the speed needed for flight: higher speeds are needed for larger animals, but higher speed means higher power requirements.

The power requirement for flying can be found from the resistive forces acting on the animal as it flies. The drag force acting on the flying animal is

$$F_L = \frac{1}{2} C_D \rho A' v^2. \tag{4.11}$$

This looks almost identical to the formula for lift force except that C_D is a different coefficient, typically about 0.1 for streamlined fliers, and A' is the effective projected forward area for the animal. This is a dissipative force, so power must be provided to keep the flier moving forward. This is an estimate; in particular, it may overestimate the power needs for larger fliers, which mostly fly by soaring, using thermals for their lift. However, it is a useful rule of thumb.

The power requirement can be found by using the formula $P = Fv$. So the power requirement for flying is given by

$$P = Fv = \frac{1}{2}C_D\rho A'v^3. \tag{4.12}$$

We are interested in how this scales as we increase the size or mass of the flier: in this formula, the two parameters of interest are the effective surface area (A') and speed (v). As seen above, $v \propto L^{1/2}$; surface area scales as L^2, so:

$$Av^3 \propto L^2 L^{3/2} \propto L^{7/2}.$$

A delightful book everyone should read is *Bird Flight Performance: A Practical Calculation Manual* by C. J. Pennycuick [191]. It is concerned with the computer modeling of avian flight. In it are formulas for estimating almost any quantity one would wish for when modeling flying creatures, on this world or any other. It is the only book I know that has a picture of a goose mounted in a wind tunnel [191, p. 51, fig. 5.6]. On p. 101 the author also mentions that metabolic rate scales as $M^{3/4}$, while flying power scales, in principle, as $M^{7/6}$, the derivation I gave above [191, p. 101]. However, his calculations are more detailed than mine, and he includes a graph of the ratio of flying power to (basal) metabolic rate derived from his computer simulations. If my very simple estimate were correct, the ratio should increase as mass increases:

$$\frac{P_f}{P_{met}} = \frac{M^{6/7}}{M^{3/4}} = M^{0.42}. \tag{4.13}$$

In fact, when plotted on a log-log graph, the slope is 0.35, reasonably close to our estimate. Whatever the true scaling, it is clear that the metabolic demands of flight increase more rapidly than metabolic rate as mass increases. This is the real reason why no massive biological fliers exist; Pennycuick estimates an upper mass limit between 15 and 20 kg for birds, which is close to the average mass of the California condor, the largest North American bird. The power required for flying is just too high for larger animals.

This puts the kibosh on a large number of mythological beasts: flying dragons, hippogriffs, wyverns, and the like seem ruled out on the grounds that their metabolism simply can't handle the power needed to fly. Perhaps this is why it is Mr. Weasley's fondest desire, in *Harry Potter and the Half-Blood Prince*, to learn what keeps airplanes up [205]. This also seems to imply that flying humanoid aliens such as the Overlords are impossible. While human-powered flight in ultralight aircraft or hang gliders is possible, the wingspans required (of order 5 m or so) dwarf the size of the people being carried, and also require elaborate means of getting up in the air: being powered by bicycle train gears, or launched from hilltops or from behind airplanes or something similar.

Poul Anderson thought up a good workaround to these problems in his stories concerning the Ythrians. These were flying intelligent aliens whose total mass was about 20–30 kg. To get around the mass restrictions, he postulated a very high metabolic rate for them: they were extreme carnivores with a special added digestive system to make their metabolism more efficient. With all of this, there were still issues with how they adapted to their environment: the population on their original home world was low because of the need for families to be separated by about 20 miles so that they could have enough hunting grounds to satisfy their enormous hunger. Before humans came to their world, they were limited to a Bronze Age culture because of their low population, which effectively forbade the growth of cities or even animal husbandry [23].[1]

When I was a child, I had a coffee-table book called, I think, *Dragons!* The authors tried to provide a rational explanation for dragon "flight" by postulating that dragons were large blimps. Dragons, the book said, generated large amounts of . . . ahem . . . methane and other gases in their digestive tracts, giving them lift. They combusted the methane when breathing fire. This is ingenious, but I have no idea whether it is possible. There may be some reason that this mechanism is fundamentally impossible, as I can't think of any animals that actually use it. I leave it as an exercise for the reader to work out the details.

4.4.2 Owl Post

One of the most pervasive aspects of life in the Harry Potter universe is the delivery of mail by owls, to the point that it is mocked by a character in Lev Grossman's book *The Magicians* [103]. Ignoring the issue of whether owls can be trained to deliver mail, it's worth considering what the upper mass limit for an owl parcel is (delivered by a single owl, that is).[2]

Anyone who has read T. H. White's *The Once and Future King* knows that owls eat mice, and that an owl can carry mice in its claws or beak [251]. So the mass of a mouse (call it 50 grams) is a lower bound. Estimating the greatest mass they can carry from first principles is probably an exercise in futility, so let's assume that the mass of the largest prey they hunt represents an upper bound. Since Hedwig is a snowy owl, let's consider them snowy owls. These large owls, with masses up to about 3 kg, will occasionally hunt hares, with mass up to perhaps 1 kg. So perhaps owl post isn't a crazy idea.

What about speed? In *Harry Potter and the Goblet of Fire*, Harry's owl Hedwig delivers messages to the fugitive Sirius Black [203]. Black is implied to be hiding out on a tropical beach somewhere. I'm going to assume that his beach hideaway is at least 1,000 km away from chilly England. Owl flight speed is somewhere around 5 m/s, or 18 km/hr [191]. I'll round up to 20 km/hr to make the math easy: to fly 1,000 km would take about 50 hours' total flight time. Assuming that the owl flew eight hours each day, this implies a total of six days' flying time to reach Black and six days to return. This is actually pretty consistent with times given in the books for exchanges of letters, so a point goes to Ms. Rowling for realism.

4.4.3 God Makes Power, but Man Makes Engines

The theme of this chapter is that animal behavior is limited by power generation. Humans as animals are capable of creating a few hundred watts of power, only a fraction of which is useful for work. Societies that run on human and animal power, common in fantasy settings, are extremely limited in scope compared to industrialized societies.

Perhaps the great distinction between fantasy and science fiction is in the nature of the societies they portray. In fantasy, magic is a substitute for technology, a means of controlling "great powers" without the need for machines. In science fiction, men work through machines to control great powers.

NOTES

1. Anderson was a master at examining how biology and physics influence and limit science fiction. He was also very knowledgeable about history and sociology, all of which play a strong role in his stories. His essay "How to Build a Planet" was a direct inspiration for this book.

2. As an aside, why do characters in the Harry Potter novels write with quill pens and ink on parchment? Try it; it's not easy. Cutting a nib from a feather is highly skilled work, and writing with one is not much less so. Quill pens are three complete revolutions behind current technology, maybe more. And parchment is an expensive substitute for paper.

PART II

SPACE TRAVEL

CHAPTER FIVE

WHY COMPUTERS GET BETTER AND CARS CAN'T (MUCH)

5.1 THE FUTURE OF TRANSPORTATION

This second section of the book deals with spaceflight, mostly manned spaceflight. Manned exploration of other planets is perhaps the most common theme in all of science fiction, with the possible exception of alien contact; the parallels of this literature with novels of the Wild West have been thoroughly explored before, and I won't go into them. Spaceflight is linked to the colonization of space, paralleling the colonization of the Americas and Africa by Europe in the eighteenth and nineteenth centuries. Space colonization has remained a persistent theme in science fiction to the present day; the website tvtropes.org refers to the 2009 movie *Avatar* as "[Disney's] *Pocahontas* . . . in SPACE!" with good reason [15].

However, on reading science fiction from the 1950s to the present, it is clear that reality hasn't matched the prediction. Robert Heinlein in *Expanded Universe* wrote that the colonization of space was as inevitable as the Sun rising in the east [121]. He also wrote in the same work that by 2000, nudism would be common and casual, and that all of us would have flying cars and automated homes. Now, I'm not making fun of Heinlein here. Well, perhaps I am poking gentle fun, but with respect: I suspect the track record for most science fiction writers in predicting the future is better than most academics, especially given that most science fiction authors, even authors of "hard" science fiction, aren't working research scientists. Predicting the future is just *hard*.

While science fiction writers aren't usually research scientists, they tend to be highly intelligent amateurs of science, often with advanced degrees in a scientific field. For example, Isaac Asimov had a doctorate in biochemistry, though he never did any research in the field, and Larry Niven has an undergraduate degree in mathematics. Gregory Benford is a physics professor at the University of California–Irvine, but he is an exception to the general rule. I want to make a very strong statement here: many of these writers know a lot more about some fields of study than professional researchers do. They are amateurs of science, meaning they are "lovers" of the fields they study. However, they are also in the game of entertaining, meaning that for the sake of a good story, they incorporate things that go against modern scientific knowledge. For example, Robert Heinlein's juvenile novel *Red Planet* is set on an inhabited Mars with thin but breathable air; even in 1949, the year the novel was published, a breathable atmosphere was known to be very unlikely. In fact, scientists had pointed this out as early as 1918 [25]. In the interests of a good story, however, it's more interesting to discuss manned colonies on Mars than unmanned robots.

Coming back to manned space travel: there are a lot of reasons why we might want manned spaceflight and colonies on other worlds. However, many of the reasons given in science fiction stories are economic at heart. The claim is that certain industries can be run more profitably in space, or that we can mine materials on the Moon or grow crops in L5 colonies or on other planets. I want to take a serious look at the economics of manned space travel and the establishment of colonies on other planets, in addition to the physics and chemistry.

Another reason given for manned space travel is military. This idea was very prevalent in the 1960s through the late 1980s when the U.S. government's Strategic Defense Initiative, more commonly known as the "Star Wars" program, was being presented as a serious space defense system. The nickname comes, of course, from the wildly popular 1977 movie *Star Wars* (now known as "Episode IV"), with its depictions of space battles looking like World War II–era dogfights. George Lucas was refighting the Battle of Midway in space, but real space battles will look nothing like this.

The final reason given is usually presented as more noble: the survival of the human race itself. This has been expressed in many different ways in the literature. Sometimes the remnants of humanity are fleeing a disaster that overtook Earth, such as nuclear war or an ecological disaster. The last story in Ray Bradbury's work *The Martian Chronicles* is of this sort, as is Larry Niven and Brenda Cooper's novel *Building Harlequin's Moon* [40][185]. Sometimes the catastrophe is more cosmological, as in Olaf Stapledon's *Last and First Men*, in which the human race moves from Earth to Venus because the Moon falls from the skies [225]. Sometimes it is more a generalized expression of the need to find a new home to safeguard the survival of humanity, in case something happens to Earth. Many stories are constructed around this sentiment. There are a number of ways in which humanity might become extinct on Earth, so it is worth exploring these ideas as well.

Two themes present themselves over and over when we examine space travel: energy and cost. The two are linked, though not as closely linked as they might at first appear. It takes a lot of energy to launch a rocket or other space vehicle from Earth to another planet, or even into orbit around Earth. It takes a lot of money to do this, too. The theme of this chapter can be summed up in the following idea: energetically, it's a lot cheaper to do a computation than to move something around, and this is why we don't yet have manned moon bases.

5.2 THE REALITY OF SPACE TRAVEL

In his nonfiction work *The Promise of Space*, Arthur C. Clarke writes that to the pioneers of space travel, Tsiolkovsky, Goddard, and Oberth, "Today's controversies between the protagonists of manned and unmanned spacecraft would have seemed . . . as pointless as the theological disputes of the Middle Ages" [55]. He was implying that unmanned and manned spaceflight would go hand-in-hand, with unmanned probes helping in the planning of larger, manned expeditions. Clarke points out that Goddard's notebooks, and most of Tsiolkovsky's and Oberth's writings, make it clear that "men would be the most important payloads which rockets carried into space." However, a lot has happened since 1968, when the book was first published, or even 1985, the date of publication of my dog-eared second edition.

What has really happened is that unmanned spaceflight, except in the popular imagination, has come to dominate the field, at least in its scientific impact. Although the Space Shuttle took up the lion's share of the NASA budget until the program's demise in 2012, its importance to science was minimal compared to that of unmanned projects such as the Mars Rovers, the Galileo missions to Jupiter, the Huygens probe of Saturn, or almost any other unmanned mission one could name. Almost the only thing of any scientific importance the shuttle did was to launch the Hubble Space Telescope and then send a mission to repair it later; important though these were, the price tag of the two launches, roughly half a billion dollars each, could have funded the building and unmanned launch of the telescope ten times over.

The contrast between perception and reality is interesting from both social and scientific perspectives. I will leave the social perspective to other authors. Much, however, of the first several chapters of this section will be devoted to the close examination of manned spaceflight and the fundamental reasons that unmanned probes have done so much more than manned ones in the past five decades. We'll consider this point carefully: it points up an interesting contrast from fundamental physics, namely, that spaceflight takes a lot of energy, whereas computing can be done essentially for free. In 1968, when Clarke wrote his book, computers were big mainframe devices that were programmed using punch cards. No science fiction writers anticipated the million-fold shrinkage in the size of computational circuits and the increase in computational speed from then till now—and we are still far from the ultimate limits of computation. The range of scientific instrumentation we can place on satellites has also increased exponentially since then, with the advent of CCD devices, x-ray and gamma-ray telescopes, and similar things. Computers can be made tiny and require little power; and they neither breathe nor eat. People, however, are large and require elaborate protection in space to keep them alive. Also, computers don't have to be retrieved from space when they have finished their mission. Let's not start with spacecraft; there are more prosaic examples to begin with. Our current transportation systems, in the form of cars, are energetically expensive when compared with computers.

5.3 THE ENERGETICS OF COMPUTATION

There's a saying that goes something like this: if cars had improved as much as computers had in the last four decades, we'd have a car that could fit into your pocket, seat 100 people, and go around the world at a thousand miles per hour on a drop of gasoline. Some of these specs are obviously contradictory, but what the hell.

Basic physics tells us why computers got so much better while cars didn't. As Richard Feynman said, there is plenty of room at the bottom. The essential operation a computer does is flip the state of a bit. A bit is a representation of a number, either 0 or 1, for use in base two arithmetic. Readers interested in details of this should read Paul Nahin's book, *The Logician and the Engineer* [173]. The archetypal computer has a memory in which bits are arranged in some way; when calculating some function, the computer can either choose to keep a given bit the same or erase it and rewrite it.

The physicist Rolf Landauer proved that the lowest energy it took to erase a bit was equal to $kT \ln 2$, where T was the temperature (in degrees Kelvin) and k is Boltzmann's constant (1.38×10^{-23} J/K) [143]. At room temperature that's only about 10^{-21} J per bit flip. Some physicists have even speculated about computing that takes no energy, but for now let's stick to this as a minimum.[1] The laws of physics also tell you that the minimum time to flip a bit from 0 to 1 or vice versa is given by

$$\tau = \frac{h}{E},$$

where E is the energy used to flip the bit and h is Planck's constant ($= 6.626 \times 10^{-34}$ J-s). In principle, if we are down to the minimum energy, the maximum speed is the inverse of this: about 10^{13} bit flips per second, or some four orders of magnitude faster than today's processors. Of course, limitations on silicon have already started to slow the pace (or even stall it), but this is the fundamental limit. The power use of this ultimate computer can be found by multiplying the energy by the frequency:

$$P = kt \ln 2 \times \frac{kT}{h} = \frac{(kT)^2 \ln 2}{h} \approx 10^{-7} \text{W}. \tag{5.1}$$

Again, we may not be able to get near this limit, but it does show why computers got so good so fast: there's a very long way to go before we approach the fundamental limits. What about automobiles and their fundamental limits?

5.4 THE ENERGETICS OF THE REGULAR AND THE FLYING CAR

Let's start with a seemingly dumb question: Why do cars need engines? This does seem pretty stupid; they need engines to move. But it's not quite so stupid when one thinks about Newton's first law: *an object in motion remains in motion*. Maybe a car needs an engine to set it in motion, but once in motion, why shouldn't it continue to go down the road forever without running the engine?

The answer to this is why Aristotelian physics retains such a hold over the mind: if there were no other forces acting on the car, it would move in a straight line forever. But there are forces, called dissipative forces, that act to slow the automobile down. Two of the forces are the friction of the tires against the roadway and the drag due to the air as the car moves through it. However, they don't account for most of the energy losses, which occur in the engine itself. A large fraction of the energy created by the engine is damped out by the motion of the pistons against atmospheric pressure and by losses in the drivetrain from the engine to the car's wheels. Beyond this, however, are the ultimate limits placed on it by the laws of physics. The earliest engines were built before the concept of energy was fully understood; Joule's experiments on the conversion of work into heat were published in 1844, fully eighty years after Watt's first commercially successful steam engine was marketed. Joule's work, and that of others, led to the *first law of thermodynamics*:

> The total energy of a closed system may be transformed from one kind to another, but cannot be created or destroyed.

This is the principle of the conservation of energy. There are a number of different types of energy; however, we are concerned with the conversion of chemical energy to work. The combustion of 1 kg of gasoline liberates 45 million J of energy; in principle, this could lift a car with mass 1,000 kg a distance of 4,500 m in the air, or roughly three miles. This is why life has become so easy since the Industrial Revolution: the amount of work an engine can do dwarfs the amount of work any individual can perform by a factor of several hundred.

However, energy conservation isn't the only issue. Not all of the energy obtained through burning fuel can go into useful work; some of it must be exhausted in the form of useless heat, randomized energy, into the environment, from which it cannot be recovered. For example, there's a lot of energy stored in a single ice cube. If the cube has a mass of 36 grams, there are about 12 trillion trillion atoms in it (12×10^{24}), and each of those atoms has about 3 kT of randomized thermal energy. (The 3 kT comes from a more complete analysis of the total energy of the system.) The total thermal energy of the ice cube works out to about

$$E = 3 \times 12 \times 10^{24} \text{ atoms} \times 300 \text{ K} \times 1.38 \times 10^{-23} \text{ J/K} \approx 150,000 \text{ J}.$$

It's not too shabby; it works out to about 5 million J/kg. The thermal energy in 1kg of ice is about the same as the amount of work one person can do in a day. So, let's solve the world's energy problems: let's take some ice, or anything else for that matter, extract its thermal energy, and use that to run our machines. Cool!

Well, that in a word sums up the problem: *cool*. Ice is cooler than a car engine. We talked about this in chapter 2 ("Harry Potter and the Great Conservation Laws") in a different context, but it's worth repeating: if we put an ice cube on top of our hot car engine, the energy from the ice cube won't flow into the engine—quite the reverse. We find that energy, in the form of heat, flows from the engine into the ice cube, warming and melting the cube while it cools off the engine. This is the *second law of thermodynamics*:

> Heat flows spontaneously from higher temperatures to lower temperatures, never the reverse.

Another formulation of it states:

> In order to make heat flow from an object at a higher temperature to one at a lower temperature, you have to do work.

If energy goes from the ice cube into the engine, it'll further cool the ice cube and further warm the hot engine, which is something that never happens on its own. If you want it to happen, you have to spend more energy to do it. In other words, you can't get something for nothing.

The second law puts stringent limits on how well any engine can work. An engine cycle is a process whereby heat generated by some chemical process is extracted from a "reservoir" at high temperature and moved to another at a lower temperature. In the process, some of the heat is used to do something useful, which is to say, it is transformed into work. For example, in the Otto cycle of a car engine, heat is generated by the combustion of gasoline and air; the reaction heats the piston to a high temperature, driving it forward via the expansion of the hot gases. When the piston returns back, it pushes the combustion products out into the air at a lower temperature; the combustion products still carry a lot of heat that couldn't be used in the work the piston did. The efficiency of the engine, defined as the ratio of the amount of useful work divided by the total energy used, has a strict upper bound:

$$e \leq 1 - \frac{T_C}{T_H}, \tag{5.2}$$

where T_H is the highest absolute temperature reached during the cycle and T_C is the lowest. Generally, the lower temperature is room temperature (around 300 K); for a typical car engine, the upper bound might be around 500 K, meaning that the upper bound on the efficiency is

$1 - (300/500) = 0.4$, or 40%. This is the theoretical limit; the actual efficiencies tend to be lower, around 20%. To be blunt: computers are ten trillion times less energy efficient than they could be, which is why they keep improving by leaps and bounds, whereas cars are within a factor of two of being as efficient as they could be, which is why improving cars is difficult.

This is an important point. The first steam engines were less than 2% efficient, but their efficiency rose rapidly, essentially on an exponentially increasing curve, until they began to approach their maximum possible efficiency. Nowadays, to improve the internal combostion engine substantially would take a lot of work. The current emphasis seems to be on hybrid electric-gasoline engine designs, but there are also plans for new materials that would allow higher maximum temperatures.

How about adding flight to the car? After all, science fiction is filled with portrayals of flying cars, from Hugo Gernsback on down. One gets into an interesting semantic problem at this point: many small airplanes aren't much larger than large cars. If we add wings to a car, does that make it a flying car or a small airplane? Or would a hovercraft count as a flying car, even though it can't get more than a few feet off the ground? In any event, let's say we have some James Bond–type car with fold-out wings: how good can we make it? Well, the airflow over the wings is what keeps the car up, but it also creates drag on the car as well. We can estimate the amount of power needed to overcome the air drag by assuming that the drag force is some fraction of the lift force on the plane—say, 20% of it. The lift force is equal to the weight of the car— if we assume a small car of about 1,000 kg, its weight is about 10,000 N; from this, there will be about a 2,000 N drag force on the plane. If the car is moving at 60 mph (which is probably too low a speed to keep it in the air), the engine power needed to overcome the drag can be estimated from the formula.

$$P = Fv, \tag{5.3}$$

where P is the power expenditure, F is the drag force, and v is the car's speed—about 30 m/s in this case. Thus the power required to overcome the drag force is around 60,000 W. However, since the car engine is only about 20% efficient in turning heat into work, it will have to operate at some 300,000 W to keep the car in the air. This is pretty hefty: it's a 400 hp engine. This is about three times the horsepower of a 2010 Toyota Corolla, which has a higher mass (and would therefore require a larger engine if it flew). So, a flying car is not impossible, but it would be expensive to maintain. And also dangerous. If you run out of fuel in a car, you pull over to the side of the road. Running out of fuel in a flying car is not something good to imagine. In addition, learning to fly a plane is much harder than learning to drive, crashes from the air are a lot scarier than crashes on the ground, and highways don't work well as landing strips if there are cars on them. I'm sure you can come up with other problems as well.

Of course, the flying taxis in most science fiction stories aren't usually portrayed as small airplanes; they're usually much more impressive, and hence much more expensive to fly. In Robert Heinlein's *The Number of the Beast*, the protagonist has a "flying" car that can be launched into a suborbital ballistic trajectory [120]. Heinlein once predicted (in 1979) that in the future (i.e., in 2010 or thereabouts), people would be able to routinely travel at thousands of miles per hour; this hasn't happened yet, alas [121]. Such travel is very akin to spaceflight, however, so we will look at it in detail.

5.5 SUBORBITAL FLIGHTS

In the next chapter I discuss vacations in space, but for the rest of this one I'll talk about Heinlein's suborbital "flying car," which is just a ballistic missile in disguise. A manned, steerable ballistic missile, but in principle nothing different from the ones that housed the nuclear missiles the United States and the Soviet Union used to point at each other. How does a ballistic missile work?

We can understand the ballistic missile (in a very rough sense) from the range equation, something all freshman physics majors are taught in the first two weeks of their first physics course. Simply put, if you throw a ball into the air at a certain speed, making a certain angle with respect to the ground, the range equation will tell you how far it travels.

There are a few caveats here:

- The equation doesn't take air resistance into consideration.
- The equation assumes that the projectile is being launched over flat ground.

These two caveats are going to restrict the validity of what I am saying here: ballistic missiles travel over large distances, so we may not be able to ignore the curvature of the Earth when dealing with them. In addition, at least the launch and landing phase of the missile take place within Earth's atmosphere, so we can't ignore air resistance in a completely realistic treatment of the problem. But the basic idea is simple: you toss the "flying car" into the air, it follows a roughly parabolic trajectory because of gravity, and it comes back down.

Here's the equation:

$$R = \frac{v^2}{g}\sin(2\theta), \tag{5.4}$$

where v is the launch speed, g is the acceleration of gravity (about $10\,\mathrm{m/s^2}$ on Earth), and θ is the angle above the horizontal that the projectile is launched. One can also work out the flight time:

$$t = \frac{R}{v\cos(\theta)} = \frac{2v\sin(\theta)}{g}. \tag{5.5}$$

The maximum travel distance R occurs for an angle of 45 degrees. Table shows the minimum speed for our ballistic flying car needed for various travel distances.

The third column is the total flight time in minutes (based on the speed and distance traveled). It isn't bad: we can travel a thousand kilometers in under ten minutes. One issue is that the formula can't be trusted for distances much greater than 1,000 km because at that point the curvature of the Earth must be taken into account. For example, the estimate of velocity for a 10,000 km hop is 10,000 m/s, which is significantly higher than the speed it would take to put the car into orbit. At this point we are into orbital flight, which is covered in the

Table 5.1
Projectile Distance as a Function of Launch Speed

Distance (km)	Speed (m/s)	Flight Time (min)	Kinetic Energy (J)
1	100	0.24	5×10^6
10	320	0.74	5×10^7
100	1,000	2.36	5×10^8
1,000	3,200	7.37	5×10^9

next chapter. The kinetic energy of the flying car is derived under the assumption that the car has a mass of 1,000 kg, which is probably too low but ballpark correct.

The fuel costs kill this idea, at least at the longest distances. Under some reasonable assumptions, the 1,000 km hop is going to run at least a few thousand dollars in fuel. This is ignoring the practical engineering stuff about how to fling someone around in a can like this and still have him or her survive the fall. That step will involve a lot of fuel as well. I don't think we'll see the flying car anytime soon.

NOTES

1. The subject of reversible computing is too complex to go into here. For a good discussion of why, in theory, computation can take no energy, see *The Feynman Lectures on Computation* for a very readable introduction to the subject [82, chap. 5, "Reversible Computation and the Thermodynamics of Computation"].

CHAPTER SIX

VACATIONS IN SPACE

6.1 THE FUTURE IN SCIENCE FICTION: CHEAP, EASY SPACE TRAVEL?

> The last conjectures may seem absurd to those who look still on manned space flight in terms of todays multimillion dollar productions. But . . . the cost and difficulty of space travel will be reduced by orders of magnitude in the decades to come.
>
> —ARTHUR C. CLARKE, *THE PROMISE OF SPACE*

In an early scene in the movie *2001: A Space Odyssey*, Dr. Heywood Floyd travels via a commercial (Pan Am) shuttle to a space station in orbit around the Earth. The humdrum way in which this is shown in the movie makes it clear that in the future world of 2001, such "flights" are common. This scene is almost unique among science fiction movies for actually getting the dynamics of space flight correct. Clarke's observations from his popular 1968 book on space exploration reinforced this idea in the movie: in the future, space travel will be as easy and as common as traveling on an airplane.

This is *the* great theme of science fiction, one that has permeated the genre since it became a separate branch of literature. Essentially every science fiction writer between 1900 and 1980 published stories involving interplanetary travel, even those writers who weren't primarily known for writing hard science fiction. A sampling of some well-known stories from this time period makes the point clear:

- *From the Earth to the Moon*, by Jules Verne. This is not the first story of space travel (Verne was anticipated by Cyrano de Bergerac and Johann Kepler), nor is it scientifically accurate: his astronauts build a cannon to launch the spacecraft, which in reality would smash them to jelly on launch [241].
- *First Men in the Moon*, by H. G. Wells. Again, this isn't scientifically accurate: Prof. Cavour invents a metal that shields anything placed above it from gravity. Even at the time, this was known to be impossible because it violated the first law of thermodynamics: one could use such a metal to build a perpetual motion machine. Regardless, the intrepid heroes build a spacecraft made from Cavourite and travel to the Moon where they meet the insect-like Selenites living in caves beneath the surface [249].
- *Ralph 124C41+*, by Hugo Gernsback. This represents the prehistory of American hard science fiction. Space travel isn't central to the theme, but the eponymous hero travels into space at the end of the novel [91].
- *Rocket Ship Galileo*, by Robert Heinlein. Heinlein is one of the best of the golden age space writers, and this book represents a wonderful blend of good writing and good science. In the book three Boy Scouts and an adult build a nuclear-powered spacecraft, anticipating the NERVA program of the 1970s. It's still worth reading, and is one of the books that drew me into science fiction in the first place. It is also worth looking at because it exemplifies what the science fiction writers paid attention to (the science) and ignored (the enormous infrastructure costs associated with projects like this). But more on this below [109].
- *The Martian Chronicles*, by Ray Bradbury. The science is unimportant to the story, but it does, of course, center on the exploration of Mars by humanity [40].
- *Ubik*, by Philip K. Dick. This is a very weird novel mostly taking place in an artificially created afterlife, but the first scenes of the novel involve a trip to the Moon. The world in which the story takes place, as in most of Dick's stories, is a near-future Earth where travel to outer space is as common as

airplane travel is today, although such details are seldom important to the story. One should note that the "near future" date of *Ubik* was sometime in the 1980s, now thirty years past [67].

- *Neuromancer*, by William Gibson. Usually thought of as the first cyber-punk novel, *Neuromancer* is firmly set in a world consistent with earlier science fiction expectations: the main characters travel to inhabited space stations in orbit around the Earth toward the end of the novel. Gibson is an interesting case because his earlier novels are firmly set in this milieu but his later ones essentially abandon these science fiction trappings [95].

I could cite hundreds of other examples. The entire genre was essentially considered the literature of space travel from about 1930 to 1980 and is still strongly linked to it in the popular imagination, especially in movies, TV, and other expressions of popular culture. One important point is that space travel is commonplace *even in stories in which the plot doesn't depend on it at all.* Of the novels mentioned above, both *Ubik* and *Neuromancer* could easily have been written without space travel. So pervasive was the idea of space travel as a commonplace of the future that both Dick and Gibson included it in their novels as a matter of course.

Accurate depictions of the science of space travel and rocketry began to appear in science fiction stories in the 1930s and 1940s; this is probably associated with the publication of Robert Goddard's massive paper, "A Method of Achieving Extreme Altitudes" in 1919 [98]. He was not the first to do serious scientific work on the subject—Tsiolkovsky in Russia and Oberth in Germany anticipated his work—but his work represents the first English-language publication on the applications of rocketry to space travel. In the article he derives a number of results, including the "rocket equation" I show below. The founding of the British Interplanetary Society in the early 1920s also increased popular interest in the subject and spurred further research on these ideas. Again, the thought was that in the near future, interplanetary travel, or at least travel into orbit, would be relatively common. This didn't happen, although in the early 2000s we began to see the beginnings of space tourism, albeit for the very rich. So, what happened? Why don't we fly to the Moon on gossamer wings today?

6.2 ORBITAL MECHANICS

We'll begin by considering the cheapest trip into space: a flight into orbit around the Earth. There have been seven people who have booked a vacation in orbit through Space Adventures, a company devoted to space tourism; Dennis Tito in 2001 supposedly paid $20 million for a ride to the International Space Station. This is clearly far beyond the reach of the ordinary Joe or Jane, but the flight was supposed to inaugurate the era of space tourism, with prices falling and larger numbers of people heading up and out. However, there has been little progress on this front in the last decade.

At first glance it would seem that the biggest problem with traveling into space is that of energy. It takes a lot of energy to get into orbit. The cost of this energy is expensive in today's society. So we need to ask two questions:

1. How much energy does it take to go into space? What are the costs of this energy usage?
2. Is this the limiting factor that makes space travel expensive? If not, what other factors are important?

6.3 HALFWAY TO ANYWHERE: THE ENERGETICS OF SPACEFLIGHT

> Once you're in orbit, you're halfway to anywhere.
> —ATTRIBUTED TO ROBERT A. HEINLEIN

There seems to be a common misconception that the late, great Space Shuttle could fly to the Moon if it had needed to. This idea was fostered by the heritage of the Apollo space program and an unfamiliarity with the scale of space. The shuttle orbit was typically 350 km (about 200 miles) above the surface of the planet, or less than the distance between Washington, D.C., and New York City. The distance from the center of the Earth to the Moon is some 380,000 km (about 238,000 miles), or about a thousand times greater than the highest shuttle orbit. If Earth were shrunk to the size of a basketball, the orbit of the shuttle would be about half an inch above the surface of the ball and the Moon would be about the size of a tennis ball, 30 feet away.

If the orbit is so close, why is it so hard to put the shuttle into orbit? You don't need huge launchers to go from Washington to New York, after all. Maybe it's because you have to go 200 miles up, instead of horizontally. This is part of the answer, but not all of it. The big reason has to do with that word "orbit." To simply launch the shuttle 200 miles straight up would take less energy than to put it into orbit (by my calculations, about one-fourth as much). However, such a rocket would rapidly fall back to Earth because it wouldn't have the speed to stay in orbit once there.

6.3.1 What Goes Up

People on the surface of the Earth don't fall through the Earth because of the solidity of the ground. The ground exerts an upward force on our feet that balances the downward force of gravity. Airplanes don't fall out of the sky because of the lift generated by the flow of air around their wings. But what keeps the shuttle up? It does have wings, but they're only for landing at the very end of a mission. In space, as we're told, no one can hear you scream because there's no air, and without air there is no lift to keep the shuttle up.

What is very interesting is that nothing keeps the shuttle up. In fact, it is continually falling toward the Earth but always missing it. The orbit of the shuttle is nearly circular: we can think of the orbit as something like a stone swing on a rope by a child at play. If the rope broke, the rock would fly off at a high speed at a right angle to the rope, following Newton's first law: if no net force acts on an object, it moves in a straight line at a constant speed. The stone flying free moves on a straight line in the direction it was just moving in before the rope broke (fig. 6.1 has a picture of the stone's path). However, the rope constrains it to move in a circle. The shuttle is just like that stone, with the rope being the attractive force of gravity between it and Earth.

If any object (the shuttle, a stone on a rope, a car turning around a curve) moves in a circle at some speed v, it must be acted on by a force

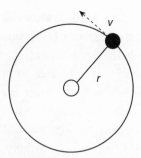

Figure 6.1. Rock whirled on a string with speed v and radius r.

directed toward the center of the circle. The size of the force is given by

$$F = \frac{Mv^2}{r},\qquad(6.1)$$

where M is the mass of the object and r is the radius of the circle. Again, the force can be anything so long as it is centrally directed ("centripetal," in physics jargon). The force on the shuttle is the force of gravity and is given by

$$F = \frac{GMm}{r^2}.\qquad(6.2)$$

Here, G is a universal constant (which has the value $6.67 \times 10^{-11}\,\text{Nm/kg}^2$ in metric units), M is the mass of the Earth ($5.98 \times 10^{24}\,\text{kg}$), and r is the distance from the shuttle to the center of the planet, which is equal to the radius of the Earth plus the height of the shuttle orbit (about 6,800 km, or 6.8×10^6 m). We can combine the two equations to solve for the speed to the keep shuttle in orbit:

$$v_{orbit} = \sqrt{\frac{GM}{r}}.\qquad(6.3)$$

One should note that the speed doesn't depend on the mass of the shuttle. Putting the numbers in, $v_{orbit} = 7,600$ m/s (7.6 km/s, or a whopping 17,000 mph). If it were any slower it would crash into the ground. Bigger speeds lead first to noncircular orbits, and eventually to the spacecraft never coming back.

The difficulty of putting the shuttle into orbit isn't mainly how high its orbit is but how fast the shuttle is moving. The payload mass of the shuttle is about 100,000 kg (roughly 100 tons). The kinetic energy of any object moving at a speed v is given by the formula

$$K = \frac{1}{2}Mv^2. \tag{6.4}$$

Using this formula, the kinetic energy of the shuttle is about 3×10^{12} J, or about 3 million MJ (megajoules). The energy density of a gallon of gasoline is roughly 100 MJ per gallon, meaning that about 30,000 gallons of gasoline contain the equivalent energy. At an average U.S. gasoline cost of $3.50 per gallon, we get a cost of about $105,000 for the amount of gasoline containing the equivalent kinetic energy. However, that isn't the entire story.

6.3.2 The Rocket Equation

The tricky part is that a rocket burns fuel to reach such a high speed. A lot of fuel. Typically, the mass of the fuel is much more than the mass of the payload for any rocket; for example, the total liftoff mass of the shuttle orbiter plus solid rockets plus tanks is about 2,000,000 kg, or twenty times the payload mass. The reason for this is that when you burn fuel to move a spaceship, you're moving not only the spaceship itself but also the mass of the fuel you are carrying. How rockets work is pretty straightforward and has to do with Newton's third law that for every action there is an equal but oppositely directed reaction. To give an analogy, let's say you are standing on a patch of frictionless ice with your little sister. You have a mass of 80 kg, she has a mass of 40 kg. She's bugging you, so you give her a shove so that she goes flying off at a speed of 4 m/s to the right. You will find yourself moving at a speed of 2 m/s to your left. Because you exerted a force on her to push her away, she automatically exerted a force on you of exactly the same size but in the opposite direction. Because you have a great mass than she does, you end up moving more slowly than she does, but you still move.

This is how a rocket works: the engine of a rocket burns fuel. The combustion of the fuel supplies the energy to move the combustion products at high speed; the engine is designed to channel the rapidly moving gases "backward"; action and reaction dictate that the ship experiences a force that moves it forward. The fuel is characterized by an ejection speed, u, which is typically a few thousand meters per second. The ejection speed is the speed at which the burned fuel is thrown from the back of the rocket. The thrust (net force pushing the rocket forward) is then given by the product of the exhaust velocity and the rate at which fuel is being burned, which we label dm/dt. We can use Newton's second law to write down a differential equation for the speed, v, of the rocket:

$$m\frac{dv}{dt} = u\frac{dm}{dt}, \tag{6.5}$$

where m is the mass of the fuel + payload of the rocket at any given time and dm/dt is the rate at which the fuel is being consumed. Two points need to be made here:

1. The acceleration of the rocket isn't constant because the mass is continually changing.
2. We are ignoring the effects of gravity on the speed of the rocket. In fact, getting to some final velocity will require more fuel than we will calculate using this equation, but the difference is relatively small in the case of the shuttle.

We can now solve the equation using elementary calculus. Assuming that the rocket started with zero speed and ended up with speed v_f, we find

$$v_f = u \ln\left(1 + \frac{m_f}{m_r}\right). \tag{6.6}$$

Here, m_f is the total fuel mass and m_r is the mass of the rocket minus its fuel. This is the famous "rocket equation," which should be memorized by any serious science fiction writer (or fan, for that matter). The shuttle booster rockets have an exhaust speed of about 2,600 m/s. Table 6.1 shows final speeds given various initial fuel mass to rocket mass ratios.

Table 6.1
Final Velocity of a Rocket as a Function of Mass Ratio

m_f/m_r	Final Velocity (m/s)*
0.1	248
0.3	682
1	1,802
3	3,604
10	6,234
20	7.915[†]
30	8,958
50	10,222
100	12,000

* The fuel ejection speed $u = 2{,}600$ m/s.
[†] Approximate orbital velocity of Earth.

Note that for large fuel ratios, the rocket can travel faster than its exhaust speed. It turns out that the shuttle needs a mass ratio of 20:1 to achieve orbital speed, according to the rocket equation. This estimate is pretty accurate: the total liftoff mass is about 2 million kg, or about twenty times the mass of the orbiter. We're off by a bit because we've assumed only one stage; that is, we've assumed that our rocket burned all of its fuel in one continuous burn, whereas the shuttle has three stages (three separate fuel burns) as it goes into orbit.[1].

Two other concepts are important for rocketry:

- Specific impulse (I_{sp}): This is just the ejection speed, u, in disguise: $I_{sp} = u/g$, where g is the acceleration of gravity. I'm not sure why rocket scientists use this instead of u, but there you are.
- Thrust (T): The net force that pushes the rocket forward, which is given by the formula

$$T = u\frac{dm}{dt}.$$

This is also essentially a materials characteristic, determined by the type of fuel you are using. Rockets are characterized by these two concepts: some

rockets have high thrust but low specific impulse, like chemical propellants; low thrust but high specific impulse, like ion rockets; or high
impulse, high thrust, such as the Orion propulsion system. Low-impulse,
low-thrust rockets are useless for spaceflight. I mostly discuss high-thrust
propulsion systems since they are the only ones capable of lifting materials
from Earth into orbit.

The total kinetic energy of the fuel burned is

$$K_{fuel} = \frac{1}{2} m_f u^2. \tag{6.7}$$

Using a mass ratio of 20, we get a net kinetic energy supplied by the fuel
of 7×10^{12} J, or about 2.5 times the initial estimate. Viewed this way, the
cost of gasoline holding the equivalent energy is about $200,000. Since
the low Earth orbit payload mass of the shuttle is about 24,000 lb, this in
principle is a cost of $8 per pound, or (since the shuttle holds seven crew
members) roughly $30,000 per person. The shuttle doesn't use gasoline,
of course, but the true fuel costs are only about ten times higher, or
roughly $2 million per launch. This would still only be a ticket cost of
roughly $300,000 per traveler. Unfortunately, the real picture isn't this
rosy.

6.3.3 The Current Cost of Space Travel

According to the NASA website, a shuttle mission has a cost of about
$450 million [8]. That is, the true cost of putting a payload into orbit
is about 200 times higher than the cost of the fuel. This is a cost of
about $19,000 per pound for the payload. I think that this is because of
the enormous infrastructure which the shuttle required: support staff,
maintenance and repair to the shuttle, processing costs for the fuel, and
so on. Typical single-use rockets have similar costs per payload pound,
although they tend to be somewhat cheaper. The Falcon-1e rocket
designed by SpaceX Corporation supposedly has the current lowest cost
per payload of about $9,000 per pound [13]. However, this rocket can
only launch satellites; it isn't capable of putting a person into orbit.

If we were to use it for space tourism, the Space Shuttle held a crew of seven. This means a cost of $64 million per person simply to break even. Of course, a lot of the space on board the shuttle was taken up by experiments. If we made a true "luxury" vehicle, maybe we could increase the number of people it would holdup to fifteen. (This is a pretty liberal estimate.) This would lower the ticket cost of a shuttle flight to $30 million.

How realistic is our estimate? Since 2001, Space Adventures has sent about seven very rich private citizens into orbit. They were launched using Soyuz rockets and stayed at the International Space Station for several days apiece. The cost was approximately U.S. $20 million, with another $15 million if participants wanted to do a space walk outside the shuttle. These numbers match pretty well our estimate of the cost per traveler using the Space Shuttle as the vehicle. When one reads about these ventures, a reason for the high infrastructure costs becomes very clear: all of the people going into space have to undergo hundreds of hours of expensive training at high-tech facilities in Russia and the United States. The personnel costs for such facilities are also high, as there are several dozen support personnel who stay on the ground for every one person going into orbit. In 2010 Space Adventures advertised a new mission: a trip around the Moon. The cost is a mere $100 million per traveler.

There's another potential cost for these vacations. Since 1986, fourteen Shuttle crewmembers died in two disasters that resulted in the destruction of the shuttle involved: the *Challenger* blew up shortly after launch on January 28, 1986, and the *Columbia* disintegrated on reentry into Earth's atmosphere on February 1, 2003. This represents two fatal disasters in 135 total flights, or a roughly one in 60 chance of being killed on any given shuttle flight. The late Richard Feynman, the Nobel laureate physicist who was on the team investigating the first shuttle disaster, harshly criticized the NASA administration for grossly underestimating the dangers of such flights:

> For a successful technology, reality must take precedence over public relations, for Nature cannot be fooled. [83]

This death rate would be absolutely unacceptable for any commercial form of transportation; the per-flight probability of being killed in an airline crash is something like one in 10 million [30]. This estimate is from a 1989 paper, but the situation hasn't changed much in the last two decades; if anything, airplane flight has become safer. For a car crash, the death rate as of 2009 was roughly one person killed per 100 million miles driven. If we assume that the "average person" drives about 15,000 miles per year, over a 20-year period this turns into a probability of about one in 166 of being killed in a car crash [16]. This means that the danger of being killed on one flight of the shuttle was about two or three times higher than the danger of being killed in 20 years of driving. Few people are going to regularly travel into space if those are the odds, even if there were a cheap way of getting there. It remains to be seen if commercial ventures such as Space Adventures are as dangerous.

Space travel needs to become much cheaper and much safer for it to become a common part of everyday life. Unfortunately, the two requirements are opposed: an increase in safety usually means an increase in cost, all other things being equal. Even something as simple as delaying a launch because of bad weather or a fuel leak can cost hundreds of thousands of dollars [14].

6.4　FINANCING SPACE TRAVEL

I'd like to summarize a few points:

- There is an irreducible minimum energy cost in putting a rocket into orbit, namely, the kinetic energy needed to get it to orbital speed.
- More energy than the minimum is required because you take the rocket fuel along with you, at least part of the way. This consideration leads to the rocket equation.
- However, the total energy costs of launching people into space seem to be small compared to the net overall launch costs, probably due to infrastructure costs.
- Over the past decade, fewer than ten people have taken self-funded orbital flights, at a cost of roughly $20 million per person.
- Finally, the danger of traveling to orbit may be unacceptably high for any sort of commercial venture. (This last item is conjectured based on the high number of people killed in the two shuttle disasters.)

The last three points are extremely important. In particular, when writing this chapter I found the point about costs very surprising. I had always assumed that the energy costs of spaceflight were the reason it wasn't more common, and that if cheap enough energy were available, space travel could become cheap. At first glance, this doesn't seem to be true. Space travel enthusiasts often allege that government bureaucracy and inefficiency kept the costs of shuttle flights high and if more commercial development of space were allowed, the costs would drop astronomically. This is an interesting point, but I am inclined to disbelieve it. Richard Feynman's book, cited above, gives a rather detailed look at the infrastructure devoted to putting the shuttle into orbit. In particular, one section stands out: he discusses the procedures the computer programmers went through to develop codes for each launch. I quote from his appendix to the official report on the *Challenger* disaster:

> The software is checked very carefully in a bottom-up fashion. First, each new line of code is checked, then sections of code or modules with special functions are verified. The scope is increased step by step until the new changes are incorporated into a complete system and checked. This complete output is considered the final product, newly released. But completely independently there is an independent verification group, that takes an adversary attitude to the software development group, and tests and verifies the software as if it were a customer of the delivered product. There is additional verification in using the new programs in simulators, etc. A discovery of an error during verification testing is considered very serious, and its origin studied very carefully to avoid such mistakes in the future [83].

This is clearly a complicated and time-consuming process: you are paying for the services of perhaps twenty highly educated professionals for several months of work for each launch. And this represents a tiny fraction of the total infrastructure needed to put the shuttle into orbit. Feynman cited this as the best aspect of the shuttle program. To bring the rest of the program to its level, presumably the infrastructure costs would need to be made even greater.

In 2006, NASA started the Commercial Orbital Transportation Services program, which is meant to fund vendors to develop and provide "commercial delivery of crew and cargo" to the International Space Station. The program is funded at a level of $500,000,000, or roughly the cost of one shuttle flight. Awards were made to Orbital Sciences and SpaceX Corporation to develop cargo delivery vehicles; the SpaceX Dragon could also potentially carry personnel to the station [12]. NASA awarded a $1.6 billion cargo delivery contract to SpaceX for deliveries of up to 20,000 kg each for twelve flights. This would represent a cost of about $6,700 per kg. Alternately, since the Dragon could be designed to carry up to seven passengers, meaning a cost of about $19 million per person carried. Unfortunately, this is not significantly different from the $20 million Dennis Tito paid for his jaunt.

From these examples, the current cost of developing a manned vehicle for near Earth orbit appears to be somewhere around $500,000,000. This sort of funding is not readily available from any other source than the federal government. Let's assume that somehow a company could develop a new reusable manned spacecraft for a cost of $100,000,000, one-fifth of our estimate; let's also assume that it had a projected life span of ten years, with ten missions per year and ten passengers per mission. Assuming fuel costs of $1,000,000 per mission and the same for infrastructure costs (which is a liberal assumption, given the discussion above), this represents a total cost of $300,000,000 over the lifetime of the spacecraft. I used a mortgage calculator on the web to calculate that if this money were borrowed from a bank at a 5% interest rate over a ten-year loan period, the developers would have to pay back $380,000,000 ultimately. Because the craft would carry 1,000 passengers over this period, each would have to pay at least $380,000 for their trip for the developers to break even.

It is possible that the reasoning used above is too pessimistic: the ultimate irreducible cost is that of fuel. As mentioned earlier, the total fuel costs were about $2,000,000 for a shuttle launch, or about $130,000 per person, assuming we could pack fifteen people into our redesigned-for-tourism shuttle. If we could bring these costs down by an order- of-magnitude (to where they are about the same as today's cost of gasoline), the fuel costs would still be about $13,000 per person (call it $10,000 if we want to round off). Let's assume that we can limit the infrastructure costs to an order of magnitude more than this value. We would then have a cost of about $130,000 for an orbital vacation (per person), at least for the ride to their "orbital hotel." This isn't within the price range of most people, but there are a few very expensive Earth-bound tourist vacations that cost about this much.

But what of this orbital hotel? Is this likely to be built also? Will we ever see structures in space capable of holding large numbers of people for settlement or visitation?

NOTES

1. Freeman Dyson showed in a very interesting paper that if one can somehow change the ejection velocity as a function of mass ratio, we can do better than the rocket equation. This is mostly of interest at high values of the mass ratio. I recommend the paper for any readers interested in the subject [72, p. 42, boxed text]

CHAPTER SEVEN

SPACE COLONIES

7.1 HABITATS IN SPACE

In the 1930s the science fiction writer George O. Smith penned a series of stories that were subsequently collected into a book titled *The Complete Venus Equilateral* [222]. The stories take place on a large space station placed in the orbit of Venus. That is, it circles the Sun in the same orbit Venus follows, but located 60 degrees ahead of Venus in orbit; it is not circling Venus. The space station serves as a radio relay, shuttling messages between Earth, Venus, and Mars, and is crewed by a few hundred engineers and support staff. The stories featured the adventures of Donald Channing, the chief engineer, and his attempts to keep the station running despite interference from mad scientists, evil bureaucrats, and space pirates. They are pure space opera and a lot of fun; additionally, the Venus Equilateral station has all the features of later proposals for real structures built in space. It has a large crew, its own ecology, and artificial "gravity," provided by rotating the structure; it is also placed in one of the Lagrange points of Venus's orbit around the Sun.

Communications satellites have taken away the need for a manned station, even if we establish colonies on other planets. However, there is still serious discussion of large manned space stations. The rationale for them can be summarized in a few points:

- They can be used as platforms in orbit for assembling deep space probes, which can be launched from there at a fraction of the cost as from Earth.
- They can be used to generate power for Earth from solar power (undiluted by Earth's atmosphere and weather, and available 24 hours a day, 365 days a year).
- They can be used for industries that benefit from a zero-G environment, such as the growth of large industrial crystals.
- They can be used for industries that require extremely high-vacuum environments.
- They could potentially be used to move heavily polluting industries off Earth and into space.
- They could be used for space tourism, as large orbital hotels.
- They represent the next step in humankind's move from Earth to inhabit the cosmos.

As in previous chapters, I'll take a hard-nosed look at these reasons. First, however, we need to look into the science behind the station.

7.2   O'NEILL COLONIES

I will first describe a community of what I like to call "moderate" size; it is larger than the first model habitat, but far below the dimensions that might be built. "Island Three" is efficient enough in the use of materials that it might be built in the early years of the next century.... Within the limits of present technology, "Island Three" could have a diameter of four miles, a length of twenty miles, and a total land area of five hundred square miles, supporting a population of several million people.

—GERARD K. O'NEILL, *THE HIGH FRONTIER*

In 1977 Gerard K. O'Neill, a physicist at Princeton University, suggested that with then current technology it was possible to build permanent habitats in space with all of the comforts of life on Earth [189]. He proposed building large space stations capable of permanent habitation for thousands or even millions of people. These structures were typically spherical or cylindrical in shape and spun on their axes to provide artificial gravity for the people living inside them.

The idea for such a structure predates him. O'Neill cites a number of predecessors, including Konstantin Tsiolkovsky and J. D. Bernal, who had similar ideas before him. The science fiction writer Arthur C. Clarke also considered Bernal an intellectual precursor, saying that Bernal's book *The World, the Flesh and the Devil* was one of the most important he had ever read. However, it was O'Neill's book that brought these ideas to the wider community, particularly as it was published shortly after the first Moon landings, when the wider human exploration of space seemed to be just around the corner.

O'Neill's motivations for building these space colonies are noble: on reading his book, one gets the sense of a liberal-minded man concerned for the future of the human race. He envisioned a time when the bulk of Earth's population would live off-planet in these habitats, when heavy industry would be moved into space to prevent it polluting the Earth, when the asteroid field and the Moon were mined for raw materials. In an article in the journal *Physics Today* he estimated that by 2008, more than 20 million people could be living off-Earth, and that by 2060 more people could be living off-Earth than on it [188]. He calculated the sticker cost of such stations to be about $30 billion for a colony housing 10,000 people (about $100 billion in 2013 money), and that such stations could pay for themselves in 20 years' time. His work led to several incredibly detailed NASA studies of large space colonies involving multiple designs. The first of them is the impressive *Space Settlements: A Design Study*, a 185-page investigation of topics as diverse as launch systems, radiation shielding, and how to plan a space-based community to avoid the psychological stresses of living in a completely artificial environment [10]. The full text of this report is available on the web, along with other studies up through 1992, the same year in which Gerard O'Neill died.

7.3 MATTERS OF GRAVITY

> The great realization was that a man falling off a roof would not feel his own weight.
>
> —ALBERT EINSTEIN

One of the many features of life in a space station, in both science fiction and real life, is that there are essentially two realistic options: either the station is essentially gravity-free or the station is spun to provide a simulation of gravity. The third option presented in science fiction stories is some form of artificial "gravity," a la *Star Trek*, but as far as anyone knows, this is impossible.

The first option, as in real life aboard the International Space Station (ISS), is that the people in the station float around in a seemingly gravity-free way. However, why they do this is subtle, as a calculation of the acceleration of gravity on the space station shows. For a spherical planet, the acceleration of gravity at any distance r away from the center of the planet is given by the formula

$$g = \frac{GM}{r^2}.$$ (7.1)

The ISS's orbit is about 400 km (4×10^5 m) above the ground. Therefore, the distance from the center of the Earth is the radius of the Earth (6,400 km) plus the height of the orbit (400 km), for a grand total of $r = 6,800$ km $= 6.8 \times 10^6$ m. The weight of an object is equal to the acceleration of gravity multiplied by its mass ($W = mg$). Because the acceleration of gravity decreases with distance from the center of the planet, we expect that someone's weight decreases as well. This is why a lot of people think that people are "weightless" on the space station: they must be far enough away from the center of the Earth that their weight drops off to nearly nothing. This is untrue. Let's take the example of an adult male with a mass of 80 kg. On the surface of the Earth, the acceleration of gravity is 9.8 m/s². Therefore, his weight is 80 kg $\times$ 9.8 m/s² $= 784$ N. (The Newton, N, is the unit of force in the metric system; it has units of kg m/s^2). If we repeat the calculation for the space station, the acceleration of gravity on board the station, using the numbers given above, is 8.7 m/s^2, meaning that the 80 kg man's weight on the station is $W' = 8.7$ m/s$^2 \times 80$ kg $= 694$ N, or about 88% of his weight on Earth. The terms "zero-G" or "microgravity" environment to describe the station are misnomers; the correct idea is that the astronauts aboard the Space Station are in free fall.

Free fall is a subtle concept. There are two ideas we need to discuss here:

- The idea that a falling person would not feel his or her own weight, and
- Why an object that is falling need not hit the ground.

7.3.1 Why a Falling Person Doesn't Feel His or Her Own Weight:

> I ... who have made the test, assure you that a cannonball that weighs one hundred pounds, or two hundred, or even more, does not anticipate by even one span the arrival on the ground ... [of one] weighing half as much when dropped from a height of 200 *braccia*.
>
> —GALILEO GALILEI, *DIALOGUES CONCERNING TWO NEW SCIENCES*

The above quotation concerns a realization by Galileo that has become fundamental in mechanics: objects acting under the force of gravity have an acceleration independent of their mass. If we take two cannonballs and drop them from a high tower, they will hit the ground at almost exactly the same time, even though their weights may differ by an enormous amount. (This is ignoring any effects that air resistance may have on them.)

Here's a thought experiment: in many amusement parks there are "drop towers", rides in which the people are lifted to some large height above the ground and then dropped toward the ground, only slowed to a stop scarily close to the ground. Now, imagine going on one of these rides with an apple in your pocket. As you are lifted up, remove the apple and then, once the fall begins, release it from your hand.

When you let it go, it falls toward the ground, of course. However, *you* are also falling to the ground with exactly the same acceleration as the apple; from your point of view, the apple simply hangs in the air. If you push on the apple, it moves away from you on what seems to be a straight-line path. This is from your point of view; to someone standing on the ground, it appears to move in a parabolic arc. You doesn't feel the force of gravity acting on you because all objects around you are falling with exactly the same acceleration. What we think of as the force of gravity isn't really the force of gravity: the heaviness we feel standing up is simply the force of the floor pushing up on us, preventing us from falling through the floor. A man falling off a roof doesn't feel his own weight, nor do astronauts in a space station that is continually falling under the influence of gravity.

7.3.2 Why a Falling Object Doesn't Have to Hit the Ground

Let's say we take a cannonball, drop it from a high tower, and measure how far it falls during the time it drops. We'll find that it falls a distance y in time t, given by the formula

$$y = \frac{1}{2}gt^2,$$

$$(7.2)$$

where $g = 9.8\,\mathrm{m/s^2}$ is the acceleration of gravity near the surface of the Earth. In 1 second the cannonball will fall a distance $y = \frac{1}{2}9.8\,\mathrm{m/s^2} \times (1\,\mathrm{s})^2 = 4.9\,\mathrm{m}$. We then try throwing the cannonball as we drop it so that our throw is perfectly horizontal—that is, the cannonball travels parallel to the surface of the Earth. Surprisingly, the distance it falls is exactly the same no matter how fast we throw it, as long as the throw is perfectly horizontal. This is because there is nothing pushing or pulling it parallel to the surface of the Earth; the only pull is toward the center of the planet. Most people have an intuitive idea that the faster the horizontal motion is, the less far it will fall in a given time, like Wiley Coyote running off a cliff in the *Road Runner* cartoons. Like most cartoon physics, this isn't the way things work.

If the Earth were flat, as it appears to be over short distances, then the cannonball would simply hit the ground after a while. But because the Earth is a sphere, the ground is dropping away a bit as the cannonball flies out. Move horizontally far enough and you'll find yourself above the surface of the Earth. How far does the cannonball have to move horizontally before it is 4.9 m above the surface? We can find this through a little geometry: let x be the distance the cannonball moves horizontally and y be the distance from the surface of the Earth, while R is the radius of the Earth (6,400 km, or 6.4×10^6 m). From the Pythagorean theorem,

$$x^2 + R^2 = (R + y)^2 = R^2 + 2Ry + y^2. \tag{7.3}$$

Eliminating R^2 on both sides of the equation, we are left with

$$x^2 = 2Ry + y^2 \approx 2Ry, \tag{7.4}$$

using the fact that y is much smaller than R. From this,

$$x = \sqrt{2Ry} = \sqrt{2 \times 4.9\text{m} \times 6.4 \times 10^6 \text{m}} = 7{,}920 \text{ m}. \tag{7.5}$$

Therefore, if the cannonball travels a horizontal distance of 7,920 m in 1 second, it will drop a distance equal to the distance that the horizon "dips." Another way of saying this is that if any projectile near the surface of the Earth is given a horizontal velocity of 7,920 m/s, it will never hit the Earth. In 1 second it will fall exactly the same distance that the Earth's surface falls away as it moves horizontally. The motion of the projectile will be circular, with a radius the same as that of the Earth.

Note that this tells us exactly the same information that equation (6.3) did; in fact, it tells us a little bit less than that equation did, because we have specified what the acceleration of gravity is. However, with a little work, the reasoning given above can be used to derive the circular velocity for a satellite at any distance from the planet.

7.4 ARTIFICIAL "GRAVITY" ON A SPACE STATION

A Space Shuttle astronaut has told me that the first few days in space feel like falling on an endlessly long roller coaster. He referred to this as the "inertia of the viscera," which results from the small displacement of one's internal organs owing to the absence of apparent weight. This is one of the medical effects of weightlessness, and astronauts who spend long periods of time in space tend to have medical problems associated with weightlessness, as the human body evolved in a world with a relatively high gravitational field.

You will find the idea in old science fiction novels, such as Robert Heinlein's *The Moon Is a Harsh Mistress*, that low gravity or weightlessness will confer longer life on humans because the heart doesn't need to pump as hard to move blood around the body [119]. This doesn't seem to be true; prolonged life in free fall leads to loss of appetite, lengthening of the body by a few inches, immune system problems, muscle and bone atrophy, and increased flatulence. None of the effects is serious in the short term, but we don't know what living in free fall for many years at a stretch will do. It's better if we can provide some sort of artificial gravity for the inhabitants of the space station.

The easiest way to do this is to spin the space station. The TV show *Babylon 5* was set on board a rotating space station five miles long in the shape of a cylinder. It rotated on its long axis to provide the sensation of gravity, as do many other space stations in science fiction. Among these are the already mentioned *Venus Equilateral*, the space station in Robert Heinlein's juvenile novel *Space Cadet*, and many, many others. Let's say that the space station is a long, hollow cylinder spinning around its (long) axis. If the cylinder radius is R and the rotational speed of the cylinder is v, the acceleration of an object moving around in the circle is

$$a = \frac{v^2}{R}. \tag{7.6}$$

The acceleration is *centripetal*, that is, directed toward the center (see appendix 1). The acceleration must be caused by a force; for a person of mass M, the force is equal to Mv^2/R and is supplied by the push of the station's hull against her feet.

Another way to look at this is to view this force from the point of view of someone inside the station, tied the rotating station. From her point of view, there is a "force" pushing her to the outside of the station: the force produces an acceleration outward from the center of the station whose magnitude is

$$g_{eff} = \frac{v^2}{r}.$$

She will attribute this to a (fictitious) "centrifugal" force pushing her outward against the hull. Therefore, a person in a rotating space station will seem to feel her normal weight if we rotate the station fast enough. If the station has a radius $R = 1$ km, we need a rotational speed

$$v = \sqrt{gR} = \sqrt{9.8\,\text{m/s}^2 \times 1,000\,\text{m}} = 99\,\text{m/s} \tag{7.7}$$

to have an "acceleration of gravity" on the station the same as at the surface of the Earth. Using these numbers, the time for one complete rotation of the station is just over one minute. However, all is not quite the same as on Earth.

7.4.1 Thrown for a Loop

There are effects resulting from the rotation of the station that are interesting and produce different results from what one would see on Earth. The rotation gives rise to two "pseudo-forces": in addition to the centrifugal force pushing against the hull, there is also a Coriolis force acting on moving objects. The Coriolis force also exists on Earth but is relatively weak compared to the force of gravity; its major action is on very large dynamical systems such as hurricanes, which it causes to spin one way north of the equator, and in the opposite direction south of it.[1] On a space station, however, the Coriolis effect is very strong because of the relatively small size of the station and the relatively large spin rate.

Imagine standing inside the station holding an apple in your hand and letting it go. From your point of view (rotating along with the station), the apple falls to the "ground" right next to your feet because of this "centrifugal" force pushing outward. However, to someone outside the station looking in, the situation is very different. Let's take two people, Susan and Mike. The names of our two observers are taken from the 1990s TV show *Babylon 5*, set in the eponymous O'Neill colony; Mike Garibaldi was the head of security for the station and Susan Ivanova was the second in command. To date, it remains one of the few TV shows that have tried to get the space science reasonably accurate, apart from having sound in space. Let's say that Susan is in a space suit floating in space outside the station, looking in. Mike is inside holding an apple; he is about to open his hand and let the apple "fall." What he sees the apple doing is very different from what she sees the apple doing. This is because if the station is large enough, Mike will not notice the rotation.

Mike and the apple are rotating along with the rest of the station. When Mike lets go of the apple, however, it doesn't rotate; instead, because of its inertia, it moves in a straight-line path tangent to the circle it was just moving on with the speed it had just before being released. This is the key point in going from one point of view to the other: the reason why Mike sees the apple fall is that his rotation along with the rest of the station means that the apple is moving. When it is free from his hand, Newton's first law takes over and it moves in a straight-line path. It hits the hull near the point where Susan's feet are rotated to as it falls. To Susan, it moves in a straight-line path at constant speed; to Mike, in the rotating spaceship, it appears to drop from his hand and accelerate to the hull. The apparent acceleration is due to the fact that he isn't moving along a straight line.

But because the apple is in Mike's hand when it is released, it is slightly closer to the axis of the space station than his feet are. It travels on a smaller circle around the center of the station than his feet do, but it travels that smaller circle in the same amount of time. Because it moves less far in the same time, it is moving with a slightly smaller speed than his feet are. If he drops it from a height of 2 m, it is traveling 0.2% more slowly than his feet. This means that it won't hit the station hull exactly where his feet are but slightly in the direction opposite the direction in which the station is spinning.

This is the Coriolis effect. (I caution that we are discussing only one aspect of the Coriolis effect here. For a complete description, see any advanced mechanics textbook, such as *Classical Mechanics*, by Herbert Goldstein [99, pp. 177–183].) If the station is rotating in a clockwise sense, Mike seems to see a force deflecting the falling apple counter clockwise as it falls. For objects that are either moving quickly inside the space station or projected up near the central axis, the trajectories are just plain odd. If you throw a ball inside the station, it seems to act under the influence of an external force. This is illusion! Once it leaves your hand, no forces act on it. From the point of view of someone outside the station looking in, it moves in a straight line. The appearance of force comes only from the rotated motion of someone inside the station. Now, even though the spin is supposed to simulate gravity, there are big differences between the trajectories of particles in the rotating reference frame and in a real gravitational field.

Because the Earth rotates, there is a Coriolis force that acts on missiles. The mathematics describing it is complicated because of the shape of the Earth and the presence of a real force (gravity) acting on the missile. A rotating space station is much simpler: the geometry is simple enough that it is very clear what is happening to the ball, but the trajectories inside the rotating frame are extremely counterintuitive.

When Michael Garibaldi throws a baseball in the garden section of Babylon 5, he seems to see two distinct forces act on it: the centrifugal force, which pushes the ball toward the hull of the station (mimicking gravity), and a Coriolis force, which doesn't appear in a true gravitational field [51][63][212]. The Coriolis force is known to be appreciable when the spacecraft diameter is small. It acts only on a moving object; the magnitude of the force is proportional to the velocity and the direction of the force is perpendicular to it. However, most science fiction authors ignore the Coriolis effect when talking about rotating spacecraft; the assumption seems to be that in large enough spacecraft, the effect will be negligible [189, pp. 122–124]. Also, the treatment of this Coriolis effect in rotating spacecraft in science fiction and speculative science has been confined entirely to the description of objects moving parallel to the spin axis of the ship; objects thrown down the long axis of an O'Neill colony, for example, appear to be deflected antispinward from the thrower. If you throw an object perpendicular to the spin axis, the general idea seems to be that you can ignore the Coriolis effect. Not so! For stations less than a kilometer or so in diameter, the effects are large and very strange. Let's look at what happens when someone in a rotating station decides to start throwing a ball around.

Consider a cylindrical spacecraft with radius R, length L. The ship rotates about the axis of the cylinder at angular frequency Ω. We set up a coordinate system x, y, z with z pointing along the cylindrical axis of the ship. In this book, we will only be concerned with motion in the xy plane. Because x and y are coordinates fixed to the ship, they do not form an inertial coordinate system. We define a coordinate system at rest with respect to the fixed stars, X, Y, a frame that is coincident with x, y at time $t = 0$. At any time t, the transformation between the two coordinate systems is given by

$$x = X\cos(\Omega t) + Y\sin(\Omega t), \tag{7.8}$$

$$y = Y\cos(\Omega t) - X\sin(\Omega t). \tag{7.9}$$

So, Mike stands inside the station at a point $x = 0$, $y = -y_0$ (see fig. 7.1) and throws a ball with velocity vector (v_x, v_y) in the x, y frame. What is the trajectory of the ball?

To Susan floating in the inertial frame X, Y, the trajectory is very simple. The ball travels on a straight line with velocity vector $(v_x - \Omega y_0, v_y)$, because once it leaves Mike's hand it is no longer under the influence of any forces. (Like all good physicists everywhere, we ignore air resistance.) To Susan, the ball travels on a straight line given by the equations

$$X(t) = (v_x - \Omega y_0)t, \tag{7.10}$$

$$Y(t) = -y_0 + v_y t. \tag{7.11}$$

To Mike, it's not at all that simple.

Let's consider a few cases:

1. The "circular orbit": $v_x = \Omega y_0$, $v_y = 0$. Mike throws the ball in an anti-spinward direction at exactly his rotation velocity. To Susan, the ball stops moving, because it has zero velocity in the inertial frame. Mike, however, is rotated under the stopped ball: he sees the ball follow a circular orbit around the center of the station. Figure 7.1 shows the trajectory of the ball

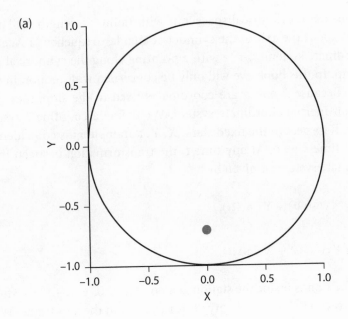

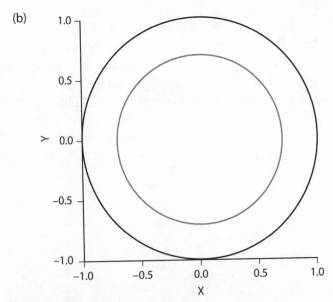

Figure 7.1. The "Circular Orbit" Trajectory.

in both the inertial and rotated reference frames. The arrow indicates the direction of the throw in the non inertial coordinate system.

This is not an unreasonable velocity for a thrown ball. Consider the cases of *Babylon 5* and another fictional spacecraft, the *Discovery*, from *2001: A Space Odyssey*. The diameter of the rotating section of *Discovery* is 11 m [58]. If *Discovery* is rotated to produce 1 g, the hull velocity is ~7 m/s. In the episode "The Fall of Night," Susan Ivanova mentions that Babylon 5's garden section rotates at 60 mph (that is, 28 m/s), which is about right, if we remember that the space station is roughly 0.8 km in diameter and the acceleration of gravity in the garden section is $1/3\,g$. Even in an Island One structure as proposed by O'Neill, the hull velocity is only about 45 m/s, the speed of a fastball thrown by a major league pitcher [189].

2. "Over the shoulder" trajectory: $v_x = \Omega y_0$, $v_y = v_x$ (fig. 7.2). Throwing the ball at an angle of 45 degrees in the air antispinward causes the ball to loop back over the thrower's head and land behind him. In the inertial frame, the ball travels along the y-axis until its trajectory intersects the hull. Because v_y is relatively large, the trajectory intersects the hull before it has enough time to make one complete rotation.

3. The "Wiley Coyote" trajectory: $v_x = \Omega y_0$, $v_y = 2\Omega y_0/\pi$ (fig. 7.3). This trajectory is so named because the ball hits the thrower in the head, from the back. In this case, the y velocity is slow enough that the station makes a complete rotation before the ball intersects the hull again. One gets the feeling that golf would be a wild and rather dangerous sport on Babylon 5.

4. The "Etch-a-Sketch" trajectory: $v_x = \Omega y_0$, $v_y = 0.08\Omega y_0$ (fig. 7.4). The smaller the y velocity, the larger the number of times the station rotates beneath it before it intersects the hull. Here is a very complicated trajectory whose initial conditions are very close to those for a circular orbit. Unlike the problem of two-body orbits under the influence of gravity, the circular orbit of figure 7.4 is unstable to small perturbations in the initial velocity.

The laws of mechanics as observed by a traveler on a small rotating spaceship or space station are so bizarre as to be cartoonlike. In the novel *Orphans of the Sky*, Robert Heinlein writes of a generation ship traveling between the stars, on a journey so long that its inhabitants forget that an outside universe exists [117]. One wonders what laws of nature that world's Galileo or Newton would invent to describe it.

(a)

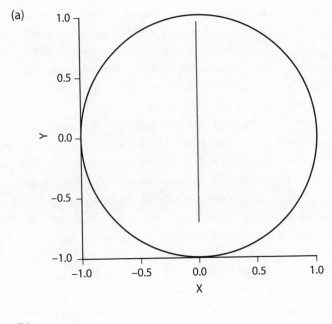

(b)

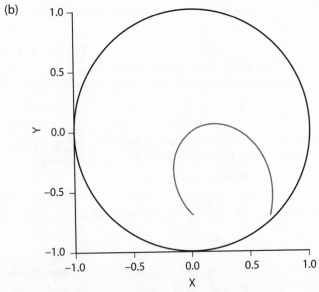

Figure 7.2. The "Over-the-Shoulder" Trajectory.

(a)

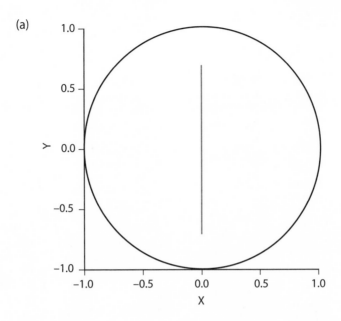

(b)

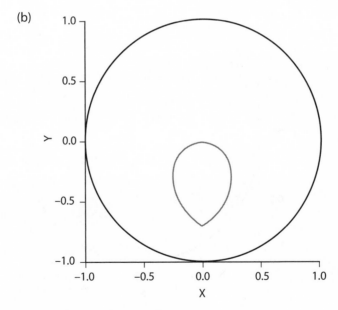

Figure 7.3. The "Wiley Coyote" Trajectory.

(a)

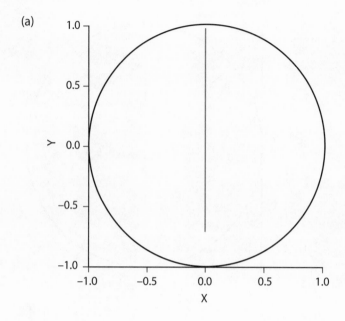

(b)

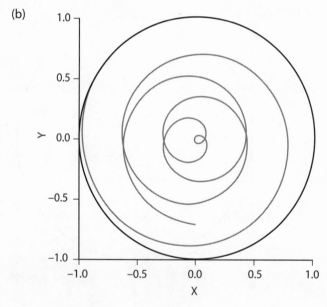

Figure 7.4. The "Etch-a-Sketch" Trajectory.

7.5 THE LAGRANGE POINTS

In previous chapters we discussed satellites in low Earth or geosynchronous orbit. These two orbits are common settings for space stations in science fiction novels, but there is a third setting that is at least as common as the other two: the L4 and L5 Lagrange points of the Earth-Moon system.

So far we've discussed satellite orbits that are solutions to the two-body problem: orbits in which only the gravitational attraction of the satellite to the Earth is important. The Lagrange points are solutions to the far more difficult three-body problem, or problems involving the gravitational interactions of three separate masses. While such problems can be solved on the computer, there are almost no solutions that lead to periodic orbits as in two-body problems.

The major exceptions are the Lagrange points, first discovered by the physicist Joseph-Louis Lagrange. These are solutions to the three-body problem under the restricted assumption that one of the three is much less massive than the other two, that is, that the orbit of the other two points can be described only by considering of their own masses and motions. The five Lagrange points are shown in figure 7.5.

The points L1 through L3 are easily enough explained by considering the balance of forces acting on them. Essentially, the force of gravity due to the Earth balances out the gravitational attraction of the Moon and the "centrifugal" force resulting from the rotation of the satellite about the Earth (or, more precisely, the center of gravity of the Earth-Moon system). However, these points are unstable: objects placed in these points will tend to drift away from them over long periods of time. They are still useful, as the time periods can be large compared to the duration of any missions. For example, the proposed orbit of the James Webb Space Telescope (the successor to the Hubble Space Telescope) is at the Earth-Moon L2 point. We can approximately calculate the position of the L1 point by balancing the force of gravity from the Earth acting on the station with that of the Moon acting on the station:

$$\frac{GM}{r^2} = \frac{Gm}{(R-r)^2}. \tag{7.12}$$

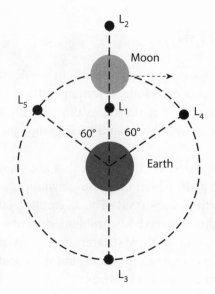

Figure 7.5. Lagrange Points of the Earth-Moon System.

In this equation,

- r is the distance from the center of the Earth to the station;
- R is the distance from the center of the Earth to the center of the Moon ($= 3.84 \times 10^8$ m);
- M is the mass of the Earth ($= 5.97 \times 10^{24}$ kg);
- m is the mass of the Moon ($= 7.35 \times 10^{22}$ kg); and
- G is the universal gravitational constant ($= 6.67 \times 10^{-11}$ Nm^2/kg^2).

One can solve this, to find

$$\frac{r}{R-r} = \sqrt{\frac{M}{m}}.$$

Because the mass of the Earth is much bigger than the mass of the Moon, the L1 point will be located much closer to the Moon than the Earth—by my calculations, nine-tenths of the distance between Earth and Moon. This is not quite all, however: there is also a *centrifugal* force, owing to the rotation of the space station around the center of the Earth, that seems to push the station away from the center of the Earth.

(It's the same pseudo force that we use to generate "gravity" on the space station.) Because it tends to offset the force of Earth's gravity (for this point) its effect is to move the position of the L1 point about 5% closer to Earth, at least by my calculations. The effect of the centrifugal force is relatively small for the L1 point but *must* be taken into account to get the position of the L2 point. The L2 point is located beyond the Moon, so the centrifugal force must balance out the gravitational attraction of both Earth and Moon. The L3 point is located on the opposite side of the Moon orbiting the Earth: again, one must balance the "centrifugal force" against the attraction of both Earth and Moon. There are a number of science fiction stories involving "counter-Earths" circling the Sun exactly opposite Earth (essentially at the Earth-Sun L3 point); we wouldn't see it because the Sun would block our view of it all the time. I *think* that this is the position of the Bizarro world of the Superman comics but cannot find a reference for this.

The L4 and L5 points are respectively 60 degrees ahead of and behind the Moon in its orbit around the Earth. They are equidistant from Earth and Moon; O'Neill proposed them as useful places for building these colonies [189]. The reason why an object that is placed there will continue to orbit is subtle; the force vector due to the resultant forces acting on it from the Moon and the Earth points toward the center of mass of the system. If the satellite's rotational speed matches the rotational speed of the Moon around the center of mass, the satellite will stay in the orbit [232].

Orbits at the L4 and L5 points are quasi-stable: if the station is pushed out of the orbit, it tends to orbit around the Lagrange point rather than be pushed away from it, as with L1 through L3 [232]. However, these points are far from both Earth and Moon, and hence energetically expensive to reach and to place satellites in. Why choose them?

His original papers show that O'Neill chose these points because he was thinking big. Remember that he projected that by the middle of the twenty-first century, more people would be living off-Earth than on it. He therefore needed room for extremely large structures, further envisioning that a network of these structures could one day be transformed into a Dyson shell, a truly huge structure surrounding the Sun that a very advanced civilization could use to harvest all the Sun's available power [188]. O'Neill also considered that mining the

Moon for raw materials would be needed to build these colonies; the Lagrange points are in some sense a compromise position between building a colony in orbit around the Earth or building one in orbit around the Moon. He pointed out, sensibly, that because escape velocity on the Moon is only about one-third the escape velocity from the Earth, the energy costs of lifting objects to the Lagrange points from the Moon is much less expensive than from the Earth. However, this requires the ability to mine the moon for raw materials, something he assumed would be possible relatively quickly post-Apollo but has not yet happened (if it ever will).

7.6   OFF-EARTH ECOLOGY AND ENERGY ISSUES

Now that we have supplied our colony with gravity and found a place in the heavens for it, what other needs do its inhabitants have? Food, air, an ambient temperature between about 0° C and 30° C: these can serve as basic human needs. However, we can't go about supplying them in the same way that we supply these needs for the Space Shuttle astronauts; this is supposed to be a self-sustaining colony, ready for long-term human habitation, not a trip lasting a few days to a week. This implies that the station needs some sort of ecology: the people on board need to grow their own food and recycle waste products so that they can live as independently of Earth as possible. Otherwise there is no way that the station can become economically viable, given the cost of transporting food and other goods into space. Building such an artificial environment will not be easy, especially when one considers the size and interdependence of ecosystems on spaceship Earth.

7.6.1   Food

An adult male with a mass of about 80 kg needs approximately 2,000 kcal of food energy to sustain him for one day; an adult female with a mass of about 70 kg needs about 1,800 kcal. (The kilocalorie, or food calorie, is a unit of energy equal to 4,200 J.) Other nutritional needs aside, this can be supplied in a variety of ways: meat is the most energy-dense food, with a caloric content of about 3,000 kcal/kg, while fruits and vegetables are less energy dense (about 600 kcal/kg). Meat, however, is very expensive in energy terms to produce, as one must raise the food animals on other food, usually some type of grain. It is about ten times more expensive to grow the grain to feed the animal, which is then slaughtered to feed the people, than it is to feed the people directly on the grain. Agricultural land on a space station will be some of the most expensive real estate anywhere, leading me to believe that very little of it will be used to raise beef, pork, or chicken. Assuming that the inhabitants are essentially vegetarian, they will need about 3 kg of grain and vegetables to eat per person per day, or about 30,000 kg/day for a station with a population of 10,000. This works out to about 10^7 kg of food per year.

At this point the choices available to the science fiction writer branch out exponentially. Will the food be grown traditionally, meaning that somehow the station must have room for acres of soil spread out and access to sunlight, or will it be grown hydroponically? Will plants be fertilized naturally or artificially? If artificially, how is the fertilizer to be produced? As an aside, the creation of ammonia via the Haber-Bosch process is very expensive, energetically speaking: modern-day agriculture depends strongly on this process, which is estimated to use some 1%–2% of all the energy used in the world [105]. Heinlein's novel *Farmer in the Sky* explores some of these issues. The novel is set on a colony on Ganymede rather than a space station. It goes into the ecology of farming in detail. The farming is done in a very low-tech manner, with topsoil being generated from Ganymedan rocks fertilized with bacteria and other organisms brought from Earth, mostly by hand and stoop labor [111]. This is an odd decision by Heinlein given the level of technology exhibited in the book, which includes limitless energy generation through the conversion of matter to antimatter.

One way to clear out these complications is by considering the conservation of energy and the idea that in growing and eating food, we are transforming energy from one form to another. At every step of the way, some fraction of the energy is lost. The prospective author must work through all the details, but let's make some assumptions and see what we get.

- First, let's assume that the space station is in orbit around the Earth or the Moon, or is in one of the Lagrange points. This is probably the most common assumption in science fiction, but it lets us assume that the flux of sunlight reaching the station is about the same as that reaching Earth from the Sun.

- Second, let's assume we have a large station to feed, and one that is somehow self-sustaining. Whether it is perfectly self-sustaining is another issue, but let's assume that most of the food consumed on the station is somehow produced on the station. Further, we'll assume that the station has some 10,000 people on board, again about the size of the station in *Babylon 5*.

- Third, let's assume that the energy to grow the crops is coming from the Sun. Again, we need not assume this, but this is the energy source of agriculture on Earth, and is free. We need not assume that we are shining the Sun directly on the growing plants; we could use solar power to generate electricity to run lights for hydroponic tanks. The solar constant above Earth's atmosphere is $1,360 \text{ W/m}^2$, meaning that every square meter of the (projected area) of the station (turned toward the Sun) is illuminated by 1,360 J of energy every second.

- Finally, let's assume a food consumption of 2,500 kcal/day per person. This is an energy usage rate; in metric units, it works out to about 120 W, or a relatively bright light bulb, per person. This means that 120 W per person $\times 10^4$ people $= 1.2 \times 10^6$ W must be available in the form of food energy.

Inefficiencies crop up in at least two places in this chain:

- The conversion of solar power into energy stored in a plant; and
- The fact that only a small portion of a plant is edible.

Plants are very roughly 1% efficient in the conversion of solar power into energy stored as biomass via photosynthesis. There are many reasons for this overall efficiency; the maximum possible efficiency is around 13% conversion from basic energetic grounds, but there are larger overall inefficiencies [245]. The second point is that only a small portion of the plant is edible; corn kernels or wheat endosperm make up only a small fraction of the overall mass of the plant, say another 1%. The overall efficiency is therefore only about $(10^{-2})^2 = 10^{-4}$. Therefore, the total power needed to supply the station for its agricultural needs is $1.2 \times 10^6\,\text{W} \times 10^4 \approx 10^{10}\,\text{W}$. Using the solar constant, the total illuminated area therefore is

$$A = 10^{10}\,\text{W}/1360\,\text{W}/\text{m}^2 = 7.35 \times 10^6\,\text{m}^2.$$

It is relatively easy to show that the other power needs for the people on board the station are small compared to the energy required for crop growth (assuming that the energy needs are comparable to that of the average American today). We can account for this by adjusting the area of the station slightly upward, say, to $10^7\,\text{m}^2$. If the station is in the shape of a cylinder 2 km in diameter, then the length of the station must therefore be about 5 km to supply these energy needs.[2] This also works out to about 1,000 m^2 per person to supply the energy needed to grow the food for their survival. A useful reference for doing these sorts of estimates in detail is the nifty little book *The Fire of Life*, by Max Kleiber. The subtitle of the book, *An Introduction to Animal Energetics*, says it all. Table 19.5 of the book is the aptly titled "Area Yielding Food Energy for One Man Per Year"; ignoring the entry on algae (presumably *Chlorella*), we see that we need between 600 m^2 to 1,500 m^2 per person, and more if we want meat and eggs. This is right in line with what I estimated [140, p. 341].

Of course, this is only an order-of-magnitude estimate, and there are other options. For example, instead of using natural lighting, one might illuminate the plants by means of white LEDs powered by large solar panels plating the outside the station or floating nearby. Even though the conversion of sunlight into electricity into the light emitted by the LEDs involves energy losses at each step of the way, one might overall do better than by using sunlight alone because much of the solar energy spectrum is useless for photosynthesis, being in the infrared spectrum. By concentrating light in the spectral region where plants can use it, the space station's inhabitants might be able to boost overall energy efficiency. This is the approach taken by Eric Yam in his design for the space station Asten. He takes a very detailed bottom-up approach, whereas the analysis in this chapter is top down, but his design has a total solar cell area of about $3 \times 10^6 \, \text{m}^2$, which is pretty close to what I calculate here [11][3].

7.6.2 Atmosphere

Have you heard about the new restaurant on the Moon? Great food, but no atmosphere.

—ANONYMOUS

Earth's atmosphere is composed of 74% N_2, 24% O_2, and roughly 2% trace gases. Earth's mean temperature is about 288 K, or 15°C, and atmospheric pressure at sea level on Earth is about $1 \times 10^5 \, \text{N/m}^2$. I'm going to assume that we need this mix of atmospheric components on our space station, although this isn't necessarily true; scuba divers use a different mix of gases (namely, oxygen and helium) when executing deep-sea dives. Oxygen is a highly reactive gas; it is not present in the atmosphere of any other planet in the solar system because it reacts with other gases or solids. On Mars, much of the oxygen is bound up in the soil or in the form of CO_2 in the atmosphere. The reason why Earth has so much oxygen in its atmosphere is life. The respiration cycle in which plants take up CO_2 and liberate oxygen, which animals then breathe and exhale CO_2, is taught in every kindergarten class in the nation. Plant life thus serves a dual use on our station, providing not only food but also oxygen for the inhabitants to breathe. This has long been realized by science fiction writers. For example, in George Smith's story "Venus Equilateral," the engineers aboard the station must circumvent disaster when the new station head misguidedly clears out the "weeds" in the station's air recycling plant, not knowing that they *are* the air recycling plant(s) [222].

We must imagine that there are enough crops or other vegetation to supply the station with its atmospheric needs. Let's do a quick, back-of-the-envelope calculation to see if we have enough plants already from agriculture to supply these needs. I'll assume the following:

- The average adult breathes about 20 times per minute.
- The volume taken in each breath is about 1 liter ($= 10^{-3}\,m^3$).
- Oxygen is generated via photosynthesis powered by the Sun at an efficiency of about 1% (as in the previous section).
- The atmosphere on the station is at the same pressure and has the same content (about 74% N_2, 24% O_2, and 2% trace gases) as Earth's atmosphere.
- There are 10,000 people on the station.

By my calculations the plants on the station will need to generate about 0.5 kg of oxygen per second to supply the needs of the people. (This is an overestimate, because not all the oxygen in each breath is taken up by the body.) The energy needs to generate this amount of oxygen can be found by considering the chemical reaction for photosynthesis [125, p. 515]:

$$6CO_2 + 6H_2O + 2,870\,\text{kJ} \rightarrow C_6H_{12}O_6 + 6O_2. \tag{7.13}$$

It's an endothermic reaction, meaning that 2,870 kJ of energy in the form of visible light must be supplied to create one mole of glucose from 6 moles of CO_2 and water. In the process, 6 moles of O_2 are formed, which can be used for human and animal respiration; since each mole of O_2 has a mass of 32 grams, oxygen can be supplied by photosynthesis at an energy cost of 1.5×10^7 J/kg. If again we assume that the overall efficiency of the process is about 1%, it will take about 7.5×10^8 W from sunlight to supply the respiration needs for the colony. This is about 6% of the energy needed for agricultural purposes, so we should be well covered by the agricultural energy budget. Another way to put it is that we need about 50 m^2 of plant area per person to provide sufficient oxygen in the atmosphere, but nearly 1,000 m^2 to provide food.[4] Interestingly enough, in the novel *Space Cadet*, Robert Heinlein uses what is clearly too low a figure for the area of plants needed to provide oxygen for a human being (10 ft^2, or about 1 m^2); the figure is off by about two orders of magnitude, as our calculations and other data show [110].

7.7   THE STICKER PRICE

> Governmental interest in high-orbital manufacturing stems in part from
> calculations on its economics. These suggest that a community in space
> could supply large amounts of energy to the Earth, and that a private,
> perhaps multinational investment in a first space habitat could be returned
> several times over in profits.
>
> —GERARD K. O'NEILL, *THE HIGH FRONTIER*

O'Neill's space stations were always meant to be economically feasible
enterprises. When counting costs he assumed, as did many others at the
time, that the dominant expense of putting people into space would be
the fuel costs. He therefore studied a number of alternative propulsion
systems such as railguns for launching materials from the Earth or
Moon. However, fuel costs are not the dominant costs for space travel
today. The dominant costs currently are infrastructure costs, which most
science fiction writers tended to grossly underestimate when writing
their books. I think that O'Neill probably fell into this trap as well; his
$30 billion investment to construct the station seems low, even when
measured in 1977 currency. Let's estimate the cost of putting a station
into space using numbers appropriate for today's space program.

Robert Heinlein's novel *Space Cadet* featured Terra Station, a large,
autonomous space colony in geosynchronous orbit around Earth. In the
book, it mentions that the mass of the station is 600,000 tons, or roughly
6×10^8 kg [110, p. 25]. From chapter 2, the current cost of putting things
into space is about $9,000/lb, or nearly $20,000/kg. Let's assume we can
lower this cost by an order of magnitude, or $2,000/kg. Then, the cost of
putting the material to make such a station would be about $1.2 trillion.

This seems about right. Eric Yam's proposal for Asten estimated that
the total cost of building it would be $500,000,000, or half a trillion
dollars, or roughly a twentieth of the GDP of the United States [11].
This figure strikes me as a low estimate. If you look at Yam's space
station budget, several items are persistently underestimated, such as the
construction processes (including mining the Moon for raw materials),
which are estimated at only $22 billion. Be that as it may, the proposal is

incredibly detailed, and I would certainly like to live on space station Asten if it ever gets built. But let's play a hard-nosed bean counter who is asked to evaluate whether the station is worth investing in. An investment of $500 billion over 20 years at 5% interest means that ultimately $790 billion will need to be paid back to investors. Since there are 10,000 people aboard the station, each of them is responsible for generating $79 million in revenue these 20 years, or about $3.2 million per year per person. This isn't impossible. A handful of the top tech companies have similar per capita revenues. Let's look at Yam's proposals for the uses of the station Asten, which are similar to ones that other space station designers have proposed:

- Mining the moon for raw materials. This is the biggest proposed use for the station. The minimum energy required to transport 1 kg of materials from geosynchronous orbit to the Moon is 4×10^6 J, as compared to 6×10^7 J from Earth to the Moon. We therefore achieve a ten times reduction in energy costs by building this satellite—more, in fact, once the rocket equation comes into the mix. One question: do we need to mine the Moon? The reason given is to build more space stations and larger spaceships, which sounds suspiciously like a circular argument to me.

- Crystal growing. This has been a mantra of space enthusiasts for the past twenty years: one can grow very large, very defect-free crystals for the computer industry in space. However, there are two problems with this proposal. First, because of improvements in crystal-growing technology on Earth, one can already grow very large, perfect crystals for the computer industry. Second, getting them from geosynchronous orbit down to Earth would cost a lot. Even if we assume something like the space elevator, the costs would still exceed $100/kg, and perhaps exceed $1,000/kg. Crystal growing is estimated as 9% of the space station's GDP; this implies that it must return a profit of at least $90 billion over the 20-year investment period we assumed, or about $4.5 billion per year. This is comparable to the yearly profits of the entire U.S. semiconductor industry, which seems a stretch for one small space station.

- Building large space telescopes and other space construction projects. Again, we can build and launch them from Earth without the enormous initial price tag.

- Other stuff. Here the Asten proposal gets plain silly. There is a discussion of creating homogeneous mixtures, "perfect spheres," biomedical research, and so forth, all of which can be done on Earth for a tiny fraction of the price tag. The author is repeating the NASA line that proponents of space station Freedom spouted in the early 1990s. They didn't make scientific sense then, they don't make scientific sense now.

The only way to build such a habitat economically is to reduce the costs of transport into space by about two orders of magnitude, to a few hundred dollars per kilogram. Let's take a look at one suggestion for doing this in the next chapter: the space elevator.

NOTES

1. The idea that water in the toilet spins the way it does because of this effect is an urban myth.

2. I'm assuming that the effective cross-sectional area for light absorption for the station is a rectangle whose length is the length of the station and whose width is equal to its diameter. This takes into account the fact that light will be absorbed obliquely because of its overall shape.

3. Asten won the 2009 Grand Prize for the NASA Space Settlement Design contest. This is a yearly contest open to high school and middle school students to design workable models for space stations. Although I am leery of whether Asten makes any financial sense, the basic physics and engineering underlying it appear to be sound.

4. Another way to do this calculation is to use the fact that in the presence of enough sunlight and a sufficiently high CO_2 concentration in the atmosphere, plants can produce a maximum of about 20×10^{-6} moles/m^2/s of leaf area, and work from there [245, p. 139, fig. 6.4]. Using this, I estimate 178 m^2 of leaf area per person needed to provide the oxygen, which is about three times larger than my back-of-the-envelope calculation. However, the moral is still the same: atmospheric oxygen needs will be provided amply by crops grown for food.

CHAPTER EIGHT

THE SPACE ELEVATOR

"... Come, let us build a city and a tower whose top may reach to heaven..."

—GENESIS 11:4

8.1 ASCENDING INTO ORBIT

It's time for your summer vacation, but you're bored with Aruba, Tokyo is so blasé, and even the wonders of Antarctica have begun to pall. So you get online and decide to book a trip to the Moon. The first leg is simple: a plane ride to Ecuador to the new Space Hub. Touching down in the airport near the Hub, you look in vain for rockets, but see only a large, circular building with what seems to be a thin ribbon reaching up as far as you can see into the sky. You go past immigration, get passports stamped to go to the R.I.L. (República Independiente de la Luna), and head into a large set of rooms the size of a train car in the shape of a ring. It has small sleeping compartments, a tiny cafeteria at the center, and an observation booth on the outside. You head over to the observation room and realize that while you were wandering around, you're now a thousand feet above the ground and climbing steadily! Three days

later, you step out into the spaceship that will take you to the Moon—
in geosynchronous orbit 36,000 km above the surface of the Earth.

The idea of a tower reaching into the heavens is as old as the Bible, and
stands as a symbol of man's ambition (and hubris) to this day. Could one
build a tower high enough to reach the heavens? Konstantin Tsiolkovsky
was perhaps the first person to investigate the physics of such a structure.
In his work, *Dreams of Earth and Sky*, he writes,

> On the tower, as one ascends it, gravity decreases gradually... due to the
> removal from the planet's center [and] also to the proportionally increasing
> centrifugal force... On climbing such a tower to a height of 34,000 versts
> [= 36,000 km] gravity will disappear, while still higher there will again appear
> a force that will increase proportionally to removal from the critical point, but
> acting in the opposite direction [26].

The idea is that if you could build a structure tall enough, by climbing up
it you could reach a point at which the centrifugal force *resulting from
the rotation of the Earth itself* perfectly balanced the force of gravity;
any object released from that point would not fall to the ground but
would remain in orbit over the surface of the Earth, orbiting once every
24 hours. Tsiolkovsky probably was illustrating the principles of orbital
mechanics rather than proposing a real structure, but real engineering
studies of such a tower began in the 1960s. Yuri Artutsanov in the USSR
and John Issacs, Hugh Bradner, and George Backus in the United States
came up with independent but similar elaborations on this basic idea and
started to look at the means of engineering such a structure [27] [128].
An elevator to take you into space may seem crazy, but there are a lot of
attractive features about it. To understand them, we need to delve into
the idea of a geosynchronous orbit.

8.2   THE PHYSICS OF GEOSYNCHRONOUS ORBITS

If you have satellite TV, you may wonder where the satellites that beam
down your programs are: how high are they above the Earth, and exactly
where do they circle? These aren't dumb questions; a lot of satellites

circle Earth in a variety of orbits. In the last chapter we found that the shuttle goes into a low Earth orbit with an orbital speed of 7 km/s. At that speed, it will circle the Earth every 90 minutes.

Any satellite in this same orbit will circle the Earth with exactly the same period, which makes it a very bad orbit for satellite TV. Such a satellite would move so fast that anyone watching it from the surface of the Earth would see it flash across the sky and vanish in a matter of minutes. The satellite receiver would be able to track it for only a minute or two every hour and a half, meaning rather intermittent coverage. We want a satellite that stays over the same spot of ground forever, meaning one that takes exactly 24 hours to rotate around the Earth.

In chapter 6 we discussed the dynamics of the Space Shuttle orbit. A geosynchronous satellite behaves in exactly the same way as the shuttle does except that it has a higher orbit. As a reminder, the speed of the satellite in orbit around the Earth is given by the formula

$$v = \sqrt{\frac{G M_E}{r}}. \tag{8.1}$$

Note that the mass of the satellite doesn't enter into this. What we really want is a relation between the *period* of the satellite and r; this relationship, which is known as Kepler's third law, is one of the most useful in all of astronomy. If T is the period of the satellite, then

$$r = \left(\frac{G M_E T^2}{4\pi^2} \right)^{1/3}. \tag{8.2}$$

Here, M_E is the mass of the Earth. One must use the correct units in this formula, which (in the metric system) for T is seconds and for M_E is kilograms. This gives an answer for r in meters. Using $T = 24$ hours $= 86,400$ s, $M_E = 5.9 \times 10^{24}$ kg and G $= 6.67 \times 10^{-11}$ N/m^2 kg^2, we find $r = 4.2 \times 10^7$ m $= 42,000$ km. This is from the center of the planet: it is approximately 35,600 km (22,000 miles) above the surface of the Earth. A satellite placed in an orbit at this distance will always stay in the same spot above the surface of the Earth as long as the orbit is around Earth's equator.

We can make the formula more straightforward if we work in nonstandard units. This is something I will do often in this book. It is sometimes awkward to always have to refer to very large numbers, like the mass of the Earth, or very small ones, like G, when doing these calculations. It is instead more useful to work in units where by the mass of the planet is expressed as a multiple of the mass of the Earth and time is measured in units of days, as these are pretty typical values for planetary masses and satellite periods. In our Solar System, the mass of the planets ranges from about one-tenth the mass of the Earth to about 300 times it. So we can rewrite the formula given above as

$$r = 42,000 \, \text{km} \times \left(\frac{M}{M_E} \right)^{1/3} \times \left(\frac{T}{1 \, \text{day}} \right)^{2/3}. \tag{8.3}$$

That is, express the mass in units of the Earth, and the orbital period in units of days; the distance is then 42,000 km times the cube root of the mass times the cube root of the square of the period.

8.3   WHAT IS A SPACE ELEVATOR, AND WHY WOULD WE WANT ONE?

Imagine taking a satellite in geosynchronous orbit and streeeetching it out, being careful to balance the upper section and lower sections so that the center always stayed in geosynchronous orbit. If we were careful, we might be able to extend one section to the ground, where we would anchor it, while the top section extended far out into space. In essence, this is the space elevator, also called a "geosynchronous skyhook" by some writers, and a "beanstalk" by others. Tsiolkovsky called it a tower, but it's really a cable hanging down from space. It doesn't crash into Earth because the center of gravity of the structure is in geosynchronous orbit and is moving at exactly the right speed to stay there. It is a truly audacious idea, but people have taken it seriously enough to fund studies on its construction [76][77].

It is a popular theme in science fiction. The first I read of such a structure was in the novel *The Fountains of Paradise*, by Arthur C. Clarke, in which the engineer Vannevar Morgan is engaged in building one [56]. The science is up-to-date: the structure is built from the carbon nanotubes discussed below. Morgan loses a fingertip to monomolecular filament built from these materials. It is a gripping story: in the climax, Morgan sacrifices himself to save several people stranded on the elevator. Charles Sheffield's novel *The Web between the Worlds* also revolves around the concept, and David Gerrold's novel *Jumping Off the Planet* involves a trip to a geosynchronous space station made using a space elevator [93][216]. The cartoonist Randall Munroe of the webcomic *xkcd* is fond of it as well.

The reason for its existence can be stated in a nutshell—energy. In principle, it costs significantly less energy to bring a person or satellite into orbit by the space elevator than by rocket. To put a satellite into orbit using rockets, you need to get the rocket moving fast on launch—7 km/s to get it into low Earth orbit, and nearly 11 km/s for a geosynchronous orbit. This takes a lot of fuel and energy. Indeed, because of the rocket equation, there is a lot of energy wasted because of the need to carry the fuel along with the rocket. These needs disappear with the space elevator. In principle, you can ascend as slowly as you want because the Earth's rotation supplies the kinetic energy needed. In fact, if the elevator extended upward above 46,000 km above the ground (52,000 km from the center of the Earth), a satellite launched from the top would escape Earth's gravitational pull completely. We'll calculate the energy needs for ascending the elevator later; they're small compared to rocketry. But how do we build such a tall structure?

8.4   WHY BUILDINGS STAND UP—OR FALL DOWN

Buildings, bridges, and other structures come in two main types: ones that are supported in compression, that is, by being pushed on, and ones that are supported partly or mostly (but not entirely) in tension — being pulled on. Building a structure depends on knowing exactly how materials respond to tension and compression: some materials are much stronger in compression than in tension, while others, such as rope or rubber bands, can take a lot of tension but can't be used at all in compression. Stone is a good example of a material that can stand a lot of compression but can't take much tension; it tends to break when pulled on strongly. European cathedrals and the Egyptian and Mesoamerican

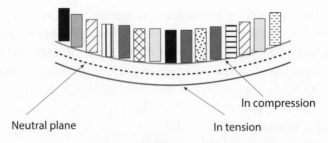

Neutral plane

In compression

In tension

Figure 8.1. Bending of a shelf loaded with books.

pyramids are good examples of structures that are supported almost entirely by compression. Such buildings tend to be much wider than they are high so that they won't bend when loaded from the sides by high wind or ground movements.

In general, structures will be in both tension and compression simultaneously. Figure 8.1 shows how this works. Consider a bookshelf piled with heavy books. The shelf bows under the weight of the books. Because of the bending, the surface immediately under the books is compressed, while the opposite side of the shelf is stretched out. Along the center of the bookshelf runs a neutral plane in which the material is being neither stretched nor compressed. From this example, you can see that bending places a structure under both tension and compression. Wood is a good example of a material that is strong in both compression and tension, which is why it is such a popular building material for small structures.

Even though there are a lot of elements in tension in a skyscraper or a suspension bridge, ultimately they are supported by the ground, or in other words, in compression. *All* structures that rest on the ground must be supported at least partly by compression. The space elevator is different: it is basically a very long cable or rope. Since ropes have no strength in compression, it must be supported entirely by tension. Exactly how it supports itself is subtle, however.

Imagine a section of the elevator that is below geosynchronous orbit—say, 200 km above the ground, the same as the shuttle orbit. As was explained in the last section, the lower the orbit, the faster a satellite must move to stay in orbit; the shuttle takes only 90 minutes to go around the Earth. However, this section of the elevator is moving more slowly than the orbital speed appropriate for that height. Left to itself, it "wants" to fall to Earth. What prevents it from doing so is the section of the space elevator immediately above it pulling it up. Newton's third law mandates that there must be a reaction force on that section pulling it down. Similarly, a section of the cable above geosynchronous orbit is moving faster than orbital speed at that height; it wants to fly away, and is held back by the section below it. Because of this, below geosynchronous orbit there is a force pulling down on the structure, while above it, there is a force pulling up. This is tension.

The tension in this structure is enormous. At the end of this chapter I include a section on calculating parameters of the structure exactly, but we can do a back-of-the-envelope calculation to estimate how strong this structure has to be. If we imagined a skyhook holding a cable dangling down above the Earth's surface, the force acting on the hook (which we will use as an approximation for the tension in the cable) is its "effective weight," which is given by

$$T = Mg_{eff} = \rho g_{eff} L A, \tag{8.4}$$

where L is the length of the structure and A its cross-sectional area. The parameter g_{eff} is the effective acceleration of gravity acting on the structure. For a short cable, this parameter would simply be the acceleration of gravity at the Earth's surface. However, the cable is so long that (as the quotation from Tsiolkovsky shows) the effective value of g will decrease as you go up the tower, reversing in direction once you pass the geostationary point. To calculate this value would be difficult, but it must lie between 0 and the actual acceleration of gravity on Earth's surface, so as a guesstimate let's choose a value about one-tenth of the actual acceleration—say, 1 m/s^2.

The parameter of interest is not the tension itself but the stress in the cable (Y), which is the tension divided by its cross-sectional area:

$$Y = g_{eff} L \rho. \tag{8.5}$$

Putting in numbers, if we estimate the cable's length as 140,000 km (see below) and the density of steel as 8, 000 kg/m^3, the maximum stress is about 10^{12} N/m^2. This estimate is three times the value of the stress indicated by an accurate calculation, but it is ballpark correct [24].

Unfortunately, the breaking stress of steel is only about $2.5 \times 10^8 \, N/m^2$, about two orders of magnitude too small. Steel clearly won't work. Is there anything that will?

8.5 STRESSES AND STRAINS: CARBON NANOTUBES

The magnitude of the problem can be seen by examining the height parameter of various materials. We can define a "breaking height" for a material by the formula

$$h = \frac{Y_{max}}{\rho g},$$

(8.6)

where Y_{max} is the maximum stress the material can take before breaking (or, more realistically, significantly deforming—what is known as the "plastic limit" for the material). Effectively, if we made a cable out of this material and hung it onto some sort of skyhook, this is the maximum length before the cable tore itself apart under its own weight. (I'm using the real value for g here, not the effective value.) For steel, h is about 3 km, or some 50,000 times less than the length the skyhook would have to be.

The ideal material to build such a structure is light (to minimize its weight) and strong. The best candidates right now are materials built from carbon nanotube fibers [207]. These are carbon compounds whose bonds are arranged in long tubes and, pound for pound, are over a hundred times stronger than steel is. Using values in the paper and a density of approximately $1 \, g/cm^3$, we calculate a height parameter $h = 1,000 \, km$. This is still not strong enough if we were to try to build a simple cable for the space elevator, but other designs make the building problems much easier.

The best design anyone has come up with for the space elevator using these fibers calls for a tapered aspect: thickest at geosynchronous orbit and thinning out both above and below [24][128][190]. These tapered structures are designed so that the stress is the same everywhere throughout the structure. The height of the structure is a whopping

150,000 km, or roughly halfway to the Moon. The advantage of this structure is obvious: the strains are minimum at geosynchronous orbit, so it makes sense to put most of the mass there. The shape of the structure is determined by the taper parameter, τ, the ratio of the thickness of the structure at the geosynchronous point to the ratio of the structure on the ground. This can be found to be [190]

$$\tau = e^{0.776R/h}. \tag{8.7}$$

Here, R is the radius of the Earth. Using $R = 6{,}400$ km and $h = 1{,}000$ km, the ratio is $e^{6.4} = 600$, meaning that the middle would need to be 600 times wider than its ends. This is not exactly impossible, but the paper I am citing uses a value of h for these nanotubes of 2,100 km, leading to a 10:1 taper ratio. Under this design, because of the lightweight nature of the carbon nanotubes, the entire structure has a mass of only 150,000 kg. This design is meant to allow a 10,000 kg payload to climb up the tower.

In principle, spacecraft launched from the top of the tower would have speeds that would let them reach Mercury or Jupiter without any additional rockets, which is another appealing aspect of the space elevator.

One issue with these carbon nanotube fibers is that as of yet, none has ever been made that is longer than a few centimeters. Stringing them into a strand 150,000 km long is a formidable engineering task, and there is a very good possibility that if this is done, the overall strength of the cable will be much weaker than the yield stress of the individual fibers.

8.6   ENERGY, "CLIMBERS," LASERS, AND PROPULSION

Another tricky issue with the tower is getting stuff up it once it is built. Current designs call for "climbers," which are essentially robots that pull themselves up along the cable. The power requirements for these climbers is pretty high, as the following estimate shows.

As far as gravity is concerned, you can think of anything sitting on the surface of the Earth as sitting at the bottom of a deep well. It takes a lot of effort to climb out of the well. The gravitational potential energy of any payload of mass m at some distance r from the center of the Earth is given by the formula

$$U(r) = -\frac{G M_E m}{r}.$$ (8.8)

The law of conservation of energy tells us that to get a climber of mass m from the surface of the Earth to geosynchronous orbit requires us to supply the difference in the potential energy. This is

$$E = \frac{G M_E m}{R_E} - \frac{G M_E m}{r_{geo}}.$$ (8.9)

We don't need to worry about the kinetic energy of our payload as the Earth's rotation will provide that through the tower; if we have a 10,000 kg payload the total energy we need to provide is 5.3×10^{11} J. To provide this power, a 2001 proposal for a space elevator suggested using a free electron laser (FEL) to beam power to optical cells attached to the climber [76]. Other ideas have been to use solar power or electrical power run up the cable. All these ideas have their drawbacks, in particular the beamed power idea. However, assuming that it can be done, let's look at the costs involved.

The first thing to realize is that there will always be some waste involved in this process. Consider the FEL idea: first, one must convert electrical power into laser radiation, which can be done with an efficiency of about 10%, give or take. Then it must be beamed to the climber, absorbed by the solar cells, and transformed again into electrical energy. The broad-wavelength solar cells that are commercially available today have a best efficiency of about 20%; because the laser operates at a single wavelength, the proper choice of materials can raise the overall conversion efficiency. The 2001 report estimated a 90% conversion efficiency, but the estimate is based on a personal communication from the developer of these cells [76]. A personal communication is not the same as a peer-reviewed publication and makes one suspicious that the quoted figure was exaggerated. Certainly there has been nothing in the scientific literature to justify this extremely high conversion efficiency. Being generous, I will assume a 60% conversion efficiency, and another

60% conversion efficiency of the electrical power into mechanical (i.e., climbing) power. This gives an overall efficiency e of

$$e = 0.1 \times 0.6 \times 0.6 = 0.036\,(3.6\%). \tag{8.10}$$

Therefore, the total energy required to bring this 10,000 kg load into orbit is $E_{tot} = 5.3 \times 10^{11}/.036 = 1.5 \times 10^{13}$ J. Assuming a cost of $0.1 per kW-hr for the electrical energy required, this is a cost of roughly $400,000, or about $40 per kilogram. This is about 500 times less than the current $20,000 per kilogram via rockets. (The report gives an overall cost of about $500 per kilogram, but this includes all costs, not just energy.) Is this too good to be true?

8.7   HOW LIKELY IS IT?

The trickiest part of building the elevator is that it is simply so incredibly long. Here the history of architecture is instructive. From ancient times until the late nineteenth century, the height of tall buildings was constrained to be less than 150 m because of the materials used (usually stone.) In 1889 the Eiffel Tower effectively doubled the maximum achievable height, and buildings have been getting taller and taller since then, with iron, steel, and concrete used in their construction. At the time of this writing, the tallest building in the world is the Burj Khalifa tower in the United Arab Emirates, which stands a whopping 818 meters (2,680 feet) high. The space elevator would be more than 150,000 times taller.

Maybe this is the wrong way to look at it. Since it's a structure in tension, maybe we should consider it as being a suspension bridge. The longest suspension bridges in the world are only about 2 km long, or 75,000 times shorter than the space elevator would be. The space elevator is so long that it could be wrapped around the world nearly four times.

Let's go through a list of things that need to be done to make the space elevator:

- Develop a climber system. This should be the easy part; space elevator enthusiasts have sponsored a $0.9 million prize for a climber to ascend a 1 km cable at a speed of 2 m/s. To date, the best entry has climbed 100 m at an average speed of 1.8 m/s.

- Develop the laser system for beamed power. Bringing a 10,000 kg climber up to geosynchronous orbit in one week requires an average power output of roughly 2.4 MW from the laser (based on the estimates provided in section 4.6). This greatly exceeds the maximum power available from any FEL built to date. By comparison, the Jefferson Labs Free Electron Laser has a maximum output power of 14 kW although there is no reason a more powerful FEL can't be built.

- Develop the carbon nanotube fibers. The strength-to-density ratio of the best fibers developed to date is still an order of magnitude less than what is needed. There are several papers indicating that long composites built from these fibers cannot achieve the strength needed because of defects. To build the cable within the next 20 years would require an average 50% increase in the yield strength per year.

- Deploy and construct the cable. The 2001 report gives a detailed analysis of how the cable could be built. First, a strand strong enough to support a climber would be deployed from a satellite initially in geosynchronous orbit; as the strand was lowered, the satellite's orbit would be boosted. Then climbers would be sent up the strand to weave in additional strands until, ultimately, it would be thick enough to support a 10^4 kg payload [76].

None of these would be impossible to achieve except perhaps developing the composite materials. The question is how much money would be needed, and how long it would take to pay off the initial costs. The big costs are the materials costs plus the deployment costs; the costs for R&D for the materials and satellite deployment system are fairly small compared to the materials and deployment costs. Both Google and the country of Japan have put money and effort into developing this idea; in 2008, Japan was willing to commit about $8 billion in funds to develop the project. Let's assume that Bradley Edward's estimate is correct and that it will take a total expenditure of $30 billion to build the entire thing. I am going to assume that the space elevator would take 20 years to build, which is what its most enthusiastic supporters contend, and that it would then have a 20-year operating lifetime. This isn't what its supporters contend, but it is reasonable given the harsh environment of space. Let's assume that the money is financed at a 7% interest rate, which is reasonable in light of the high risk of the venture. A mortgage calculator on the web tells me that financing this venture over the 40-year period would cost roughly $90 billion.

At the end of his phase 2 report, Edwards states that the total capacity of the system would be putting 1,000 tons of payload into space per year, or about 10^6 kg [77]. This strikes me as a very optimistic estimate: because each payload is about 10^4 kg, this means running 100 trips per year, or once every three days. I'll assume an order of magnitude less than this: ten payloads per year, or 10^5 kg into space per year. In a 20-year working lifetime the space elevator could lift about 2×10^6 kg (roughly five million pounds) of payload into space at a cost of roughly $1 billion above construction and development costs, assuming the canonical $250 per pound payload costs of the elevator. The total investment in the space elevator would therefore be about $91 billion, including interest, or the equivalent costs of lifting about two million pounds of payload into space by conventional means. Over the allotted time period we get a roughly 2.5-fold savings in cost in launching satellites into geosynchronous orbit compared to conventional means, if this estimate, which was a very favorable one, is correct. This is decent, but not the 99% savings that some enthusiasts offer. Again, this figure is reached by assuming a much lower capacity than the authors of the study do.

Bottom line: the space elevator is a very risky investment that offers moderate savings over conventional means of space transportation. Lowering infrastructure costs for conventional space systems might make them even more competitive with the space elevator. I've read the statement that people will build the space elevator 20 years after everyone stops laughing at it. Well, no one is laughing anymore, but I think the case has still to be made for it.

8.8 THE UNAPPROXIMATED ELEVATOR

This section is for the math and physics wonks who want to repeat the calculations in this chapter in a more exact way. The work here is based on P. K. Aravind's paper in the *American Journal of Physics* [24].

The analysis of this structure is based on the technique of free-body diagrams, which can be found in any standard elementary physics textbook, such as *Fundamentals of Physics* by Halliday, Resnick, and Walker. In a free-body analysis, we analyze and resolve the forces acting on one element of the structure independent of the rest of it; because the structure is in equilibrium, the net force acting on any independent element must be zero.

Let's consider a section of the elevator at distance r from the center of the Earth and of length dr. The tension in the structure at distance r is $T(r)$. There are four forces acting on the structure: the tension at the top and bottom, tending to pull the structure apart; the weight of the structure, pulling it down; and a "centrifugal force" due to the rotation of the Earth, tending to push it outward. We can write this as

$$
\begin{aligned}
F_{net} &= T(r + dr) - T(r) + F_c - W \\
&= T(r + dr) - T(r) + \Omega^2 r \rho A(r) dr - \frac{GM}{r^2}\rho A(r)dr \\
&= 0.
\end{aligned}
\tag{8.11}
$$

Here, A is the cross-sectional area of the structure (which may depend on height), ρ is the density of the structure, M is the mass of the Earth, G is the universal gravitational constant, and Ω is the angular frequency of the Earth's rotation ($= 2\pi/\text{day} = 7.3 \times 10^{-5}/\text{s}$). If we make dr infinitesimal in length, we can rewrite the equation as

$$
\frac{dT}{dr} = A(r)\rho\left(\frac{GM}{r^2} - \Omega^2 r\right).
\tag{8.12}
$$

We can think about solving this in two different ways:

1. By specifying the cross-sectional area as a function of r. The simplest is cross-section area A is constant.
2. By specifying some condition on the stress of the elevator, for example, making the stress throughout the structure constant.

I'll solve the first case, specifying the cross-sectional area as constant, here. I leave the second problem as an online exercise.

The stress, Y, is the tension divided by the cross-sectional area: $Y = T/A$. Therefore, we can rewrite the equation

$$\frac{dY}{dr} = \left(\frac{GM}{r^2} - \Omega^2 r\right)\rho.$$

The condition at the ground is that $Y(R) = 0$, where R is the radius of the Earth. The equation can be solved by integration:

$$Y(r) = \left(\left(\frac{GM}{R} - \frac{GM}{r}\right) - \frac{1}{2}\left(\Omega^2 r^2 - \Omega^2 R^2\right)\right)\rho. \tag{8.13}$$

The maximum value of the tension occurs at the radius of the geosynchronous orbit:

$$r_{geo} = \left(\frac{GM}{\Omega^2}\right)^{1/3} = 42{,}000 \text{ km}. \tag{8.14}$$

The maximum value of the stress is

$$Y_{max} = \left(\frac{GM}{R} + \frac{1}{2}\Omega^2 R^2 - \frac{3}{2}(GM\Omega)^{2/3}\right)\rho. \tag{8.15}$$

Using parameters appropriate for the Earth and a density of 8,000 kg/m^3 (appropriate for steel), then the maximum stress is 1.2×10^{13} N/m^2.

CHAPTER NINE

MANNED INTERPLANETARY TRAVEL

9.1 IT'S NOT AN OCEAN VOYAGE OR A PLANE RIDE

In Robert Heinlein's novel *Podkayne of Mars*, the eponymous heroine takes a voyage from her home on Mars colony along the first leg of a projected three-legged voyage to Venus, Earth, and back home again [118]. The ship she travels on, the *Tricorn*, is essentially an ocean liner moved into space; it has fancy dining halls, a ballroom, and first-class, second-class, and steerage compartments. The ship is described in detail. As is customary in a Heinlein novel, the science is pretty good: the one place where it departs most from reality is the "torch" drive, which is essentially direct matter-energy conversion, allowing the ship to travel under its own acceleration for a good part of the voyage. Even so, the trip to Venus alone takes several months because of the huge distances involved.

Other novels are filled with similar voyages. Many of them depict long voyages in cramped quarters, and in some the plots are driven by the ennui that results. In Arthur C. Clarke's *2001: A Space Odyssey*, three of the five crewmen aboard the *Discovery* are placed in suspended animation because the limited life support aboard the ship couldn't handle more than two crewmen for the multiyear voyage (to Saturn's moons in the novel, Jupiter's in the movie). So: trips to low Earth orbit take a few hours; a trip to the Moon takes a few days; and trips to even the nearer planets take months to years, even in science fiction novels, at least in scientifically accurate ones. Why is that?

Well, this is a combination of two related factors, distance and gravity. At its closest approach, Venus, the nearest planet to Earth, is about 40 million km from the Earth, or 110 times farther away than the Moon. Mars is even farther away: at the closest approach, it is 78 million km from Earth. The Apollo astronauts traveled to the Moon in three days; at the same average speed, it would have taken them almost a year to reach Venus.

The other tricky part, gravity, is due to the fact that all of the planets in the Solar System are in orbit around the Sun; they must follow the laws of gravity as they orbit it, as must any spacecraft launched from Earth to another planet. While the spacecraft is traveling from Earth to Mars, it is essentially a tiny planet in orbit around the Sun. This means it can't simply be pointed at Mars and launched off: Mars is moving in its orbit. Even if the craft could somehow travel on a straight line, when it got there Mars would be somewhere else. And the rocket doesn't move on a straight line: as I said, it's essentially a little planet too.

9.2 KEPLER'S THREE LAWS

We need to understand how planets orbit the Sun. Johannes Kepler postulated three laws governing orbital dynamics. He presented them as empirical laws, but Isaac Newton showed later that each of the laws was a consequence of his (Newton's) law of universal gravitation. The derivation, which is complicated, may be found in any textbook on orbital dynamic, such as *Fundamentals of Astrodynamics* [31].

Kepler's first law. The shape of the orbit is an ellipse, with the Sun at one focus. This is surprising to some people, who imagine that the orbits of planets around the Sun are perfect circles. In fact, the "perfection" of the circular shape was so powerful a notion that for several thousand years most astronomers felt the orbits *had* to be circles, or at least circles traveling on circles. An ellipse is the shape you get when you place

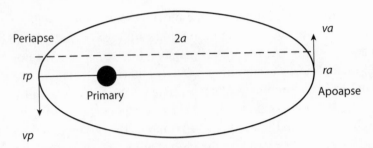

Figure 9.1. Ellipse geometry.

two thumbtacks into a board, tie a string between them so that the string is slack, and, taking a pencil, pull the string taught and trace out the resultant figure on the board (fig. 9.1). It looks like a squashed circle. The foci of the ellipse are the points where the thumbtacks are. Interestingly, there is no easy way to get this result. Richard Feynman presented a lecture in which he showed how to derive the result without using calculus, but his geometrical derivation, while "elementary," is much more difficult than standard ones using differential equations [101]. Another interesting thing is that while the Sun is at one focus, there's nothing particular at the other one.

Three useful definitions: half the length of the ellipse along the long way is called the *semimajor axis* (designated a), half the short way is the *semiminor axis* (b), and the eccentricity, e, is defined as

$$e = \sqrt{1 - \left(\frac{b}{a}\right)^2}. \tag{9.1}$$

The larger the eccentricity, the more squashed the ellipse is. As you might guess, an ellipse with $e = 0$ is a circle.

Kepler's second law. If we draw a line segment between the planet and the Sun, and let the segment sweep out a sector, the rate at which the area of the sector increases is constant. Another way to put this is, equal areas are swept by the line segment in equal times. This is illustrated in figure 9.2. The law of areas tells us that the planet moves faster when closer to the Sun, as the arc swept out by the line segment in a given amount of time has to be bigger than when the planet is farther away, to sweep out the same area.

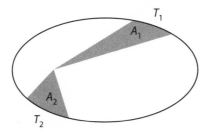

Figure 9.2. Kepler's second law: If time T_1 = time T_2, then area A_1 = area A_2.

Kepler's third law. We've seen Kepler's third law before in chapters 6 and 8. I am now going to present this law in the form that astronomers use, because it is much simpler than how physicists present it. Essentially, this involves using the right units to make the problem simpler. I did this once before in chapter 8, where I simplified the formula for the distance of a circular orbit in terms of the period of the orbit and the mass of the planet it is circling. In that case I used a period measured in units of days and a mass measured in units of the mass of the Earth. Because planets circle the Sun, I am now going to use units of period of years because planets take years to circle the Sun. I'm going to measure masses in units where the Sun's mass (1.99×10^{30} kg) is 1, because the stars have masses around the same as the Sun's. I'm also not presenting it as Kepler understood it but as it is derived from Newton's law of universal gravitation.

Let's start by considering a planet in an elliptical orbit around the Sun. It has some average distance from the Sun, a, which can be shown to be equal to the semimajor axis of the ellipse. We are going to measure the distance, a, in astronomical units (AU): 1 AU is the average distance from the Earth to the Sun (1.5×10^8 km, or 93 million miles). It takes a certain amount of time to go around the Sun; call that its period, P. We'll measure the period in years; thus, the average distance and period for Earth are $a = 1$ AU and $P = 1$ year. For Mars, they are $a = 1.52$ AU and $P = 1.88$ years.

The final thing we need is the mass of the star, the object these satellites, the planets, and our spacecraft are in orbit around. Call this M: we'll measure this in units in which the mass of our own Sun is equal to 1. I do all this because using these units makes the formula easier.

Converting to metric units from this point is then easy, and can be done in one step. Using these, Kepler's third law states that for all satellites orbiting a primary with mass M,

$$\frac{a^3}{P^2} = M. \tag{9.2}$$

If we have two objects whose masses are roughly equal in orbit around each other, we have to modify the formula slightly (we'll come to this later on).

For planets (or any other objects) circling the Sun under the influence of gravity,

$$a^3 = P^2, \tag{9.3}$$

because $M = 1$ in these units. This is almost all we need to know to plot simple trajectories for space voyages to other worlds.

9.3 THE HOHMANN TRANSFER ORBIT

Let's imagine a trip from Earth to Mars. A realistic space voyage is not much like what is shown on TV or in the movies: most movies or TV shows picture the spacecraft with their engines turned on all the time, like a plane or boat. This is not necessary and in fact wastes fuel and energy: airplanes have their engines on all the time because of air resistance. If the engine were turned off, the plane would crash in a short time.

The same isn't true in space. In a vacuum the resistance is so low that the biggest problem with the Space Shuttle was getting it out of orbit when it had to return to Earth! Otherwise it would have continued circling the planet for years or centuries. This is even more true in interplanetary space, where the vacuum is even greater: a spacecraft placed in orbit around the Sun could conceivably circle it for billions of years, just as the planets do.[1]

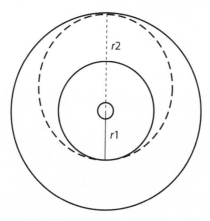

Figure 9.3. Hohmann transfer orbit.

To get a spacecraft from Earth to Mars, we want to put it into an orbit around the Sun that intersects the orbits of Earth and Mars. The simplest example of such is a Hohmann transfer, or cotangent, orbit. It was first thought up by Walter Hohmann in his 1925 work, *The Attainability of Heavenly Bodies*; the 1961 NASA translation is available free on the web [124]. I'll approximate the orbits of Mars and Earth as perfect circles with radii 1.52 and 1.00 AU, respectively. I'm also going to assume that the orbits lie in the same plane. Neither of these assumptions is correct, but both are close. The eccentricity of each orbit is .0167 and 0.0935, and the orbit of Mars is tilted by only about 1.85 degrees from Earth's. (All astronomical data are taken from appendix 4 of the textbook *21st Century Astronomy* [130]). To get the spacecraft from Earth to Mars, we place it in an orbit that is tangent to Earth's orbit at its closest approach to the Sun (its perihelion) and tangent to Mars's orbit at its farthest distance from the Sun (the aphelion). Figure 9.3 shows the geometry of the Hohmann orbit.

We'll get into how to do this in the next section. For now, what you need to realize is that we now know almost anything we could want to know about the orbit. For example, the trip time is half of the orbital period, which we can figure out from Kepler's third law: the semimajor axis of the ellipse is the average of the two radii, or 1.26 AU. The period is therefore $P = 1.26^{3/2} = 1.4$ years, meaning the total trip time to Mars is 0.7 years, or 255 days.

9.4   DELTA v AND ALL THAT

The spacecraft doesn't use its engines for most of the trip. For most of the trip it acts as a satellite of the Sun, moving under the influence of gravity. In principle, there are only two times during the outward voyage where it burns fuel: when it leaves Earth's orbit and when it enters the orbit of Mars. The reason is that the spacecraft needs to be moving faster than Earth does around the Sun to get into the orbit, and when it reaches Mars it is moving more slowly than Mars does in its orbit. At each of these points, the spacecraft needs to burn fuel to change its velocity. This leads us to the concept of "delta v" (Δv): the change in velocity needed to make an orbital correction.

To calculate this we need to get the speed of the two spacecraft at perihelion and aphelion. This can be calculated by using two ideas, the conservation of energy and the conservation of angular momentum. I will spare you the details of the calculation. At perihelion, the spacecraft's velocity is

$$v_p = \sqrt{\frac{2G\,M_{sun}}{r_p}\,\frac{1}{1+r_p/r_a}}, \tag{9.4}$$

and at aphelion its velocity is

$$v_a = \sqrt{\frac{2G\,M_{sun}}{r_a}\,\frac{1}{1+r_a/r_p}}. \tag{9.5}$$

Here, M_s is the mass of the Sun, r_p is the distance from the Sun at perihelion, and r_a is the distance of the Sun at aphelion. One problem: we have to express everything in metric units to get this to work out correctly. (Note, however, that this will work for any planet in any orbit circling any star.) Putting the numbers in, $v_p = 32,670\,\text{m/s}$ and $v_a = 21,495\,\text{m/s}$. The velocity at perihelion is greater than the velocity of Earth in its orbit (29,7500 m/s) and the velocity at aphelion is less than the velocity of Mars in its orbit (24,100 m/s). These are the two times when we need to burn rocket fuel: Δv_p represents the change in velocity at perihelion to get the rocket into the Hohmann orbit and Δv_a

represents the change in the velocity at aphelion to get the rocket into the orbit of Mars. $\Delta v_p = 2{,}925$ m/s and $\Delta v_a = 2{,}633$ m/s.

This isn't so bad at first glance; the Δv values are smaller than the speeds needed to get from the ground into near Earth orbit (roughly 7,000 m/s). In fact, the two values are about equal to the exhaust speed of a typical chemical propellant; the rocket equation from chapter 6 tells us that we need a mass ratio of about 2.8 for a spacecraft to reach this speed. I'm going to round this up to 3 to make calculations easier. Therefore, for a 10,000 kg payload spacecraft, we need a total mass of about 30,000 kg. That is, if we consider only *one* orbital maneuver.

The tricky part is that we have to perform this maneuver four times for a manned spacecraft, resulting in two Δv's going out and two coming back. The payload ratios are multiplicative. If the ratio is 3 for one maneuver, it becomes $3^2 = 9$ for two, $3^3 = 27$ for three, and so on. If we don't assume that the crew can find fuel on Mars (more on this below), the net payload ratio is therefore $3^4 = 81$. This can be done. A 10,000 kg payload requires a total spacecraft mass of 810,000 kg. But it gets trickier when we consider the launch schedule.

One other issue is that I have assumed that the spacecraft started from an orbit that was the same as Earth's orbit around the Sun; however, to get into that orbit, it essentially must free itself of Earth's gravitational attraction, which in itself is a fairly high Δv maneuver, about 11 km/s. I'm ignoring this for the time being, but it and any Δv's needed to move the craft into orbit around Mars must be added into the entire Δv budget.[2]

9.5 GETTING BACK

It's pretty obvious that the planets must be aligned in a precise way to send this spacecraft out. To get the orbits to work out, we need to start at a time when Mars will be in the correct place *at the end of the trip*. That is, at the end of the trip, Mars must be opposite, with respect to the Sun, where Earth was at the trip's start.

The math behind this is a little finicky, but I calculate that such proper alignments happen once every 26 months. You don't have to use a cotangent orbit, of course, but it makes sense to try to use the one that requires minimum energy. However, once at Mars you have to wait around for a while for the planets to line up again to go back. This means that the crew not only has to survive for 255 days going out but must endure more than a year waiting around for the orbits to be right again, and spend another 255 days coming back. In other words, they must while away nearly three years either in space or on the surface of Mars. You see right away why it is much simpler to send out unmanned probes. We don't need to worry about keeping them alive or getting them back. Probes like the Mars Rovers are much smaller, much lighter, and much more fuel efficient than a manned probe, and so are much, much cheaper. A manned Mars mission would cost more than $10 billion, possibly more than $100 billion, because of the need for the crew to return to Earth alive. Of course, there have been recent suggestions that we send out people who would simply stay on Mars for the rest of their lives, mooting a return trip. At the present level of technology, this is unfeasible. It is simply a slow form of suicide for the "colonists" involved.

9.6   GRAVITATIONAL SLINGSHOTS AND CHAOTIC ORBITS

I said in the last section that the Hohmann transfer orbit is the minimum-cost orbit, but that isn't exactly true. In reality, it is the minimum-cost orbit only if we ignore the gravitational fields of the planets the spacecraft is traveling between and consider only how the gravitational field of the Sun influences the spacecraft. This simplification is what is known in physics as a two-body problem; the only bodies being considered are the Sun and the spaceship, so the spaceship travels on a nice elliptical orbit, as Kepler's laws demand. However, Kepler's laws work only if we can ignore the gravitational influence of the planets along the trajectory of the spacecraft.

In chapter 7 I discussed one solution to the three-body problem, the Lagrange points between Earth and Moon. These represent periodic orbits of the system. There's another way one can make use of a third planetary body, however: to change the orbit of a spacecraft by a close maneuver to the planet. This is known as a *gravitational slingshot* because one can use the gravitational field of the planet to change not only the direction of the spacecraft but also its speed. It can be used to save fuel costs for long interplanetary voyages. The most famous uses of such slingshot orbits were for the Voyager probes, which were launched in the late 1970s; the probes looked like pinballs as they seemed to career off the outer planets. The purpose for using the gravitational slingshot was to visit as many of the outer planets as possible using the two probes. One side effect of the mission was that the probes were left with a speed greater than the escape velocity of the Solar System; they are true interstellar voyagers, if slow ones.

Here's the basic idea. As the spacecraft goes around the Sun, we arrange it so that its orbit passes close by (within a few million miles) of one of the planets or planetoids around the sun.[3] It uses the gravitational field of the planet to change the direction it is heading in and the speed it is traveling at. The maneuver takes place over a short time, short enough that we can consider only the motion of the planet and the spacecraft as it goes around it. We'll ignore the gravitational attraction of the Sun so that we can approximate the motion of the planet and spacecraft as being on straight lines. In science fiction, an interesting example of this occurs in Larry Niven's novel *Protector*, in which the Brennan-monster uses the gravitational field of a neutron star to make a right-angle turn in space; it is interesting because the direction change is made at relativistic speeds, which is a complication I will not go into here.

I'll do only the simplest case, and leave a more complicated example for the problems. It's remarkable how little information we need to calculate the final speed of the spacecraft. In the simple example, we will assume that initially the spacecraft is moving in the opposite direction as the planet, makes a 180-degree turn around it, and ends up moving in the other direction. If the planet is moving with speed V and the spacecraft is initially moving with speed v, after it goes around the planet the spacecraft will be moving with a new speed, $v' = v + 2V$. This comes from the conservation of momentum and energy and is independent of the details of the maneuver. All we need to know is the speeds and the initial and final directions of the spacecraft with respect to the motion of the planet.

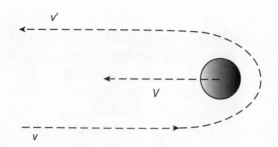

Figure 9.4. Slingshot maneuver.

The neat thing about this is that if the planet is moving in a circular orbit, its speed is 70% of the escape speed of the Solar System at that distance from the Sun. Therefore, the new speed of the spacecraft will be greater than the speed needed to completely escape from the system. In principle, we can use maneuvers like this to get a spacecraft anywhere in the Solar System. The issue, of course, is that the planet has to be in the right place at the right time to be able to do this; the Voyager spacecraft were launched during a window of time in which one could do several of these maneuvers on a single voyage, but such alignments of the planets are rare.

The third body of choice on a mission from Earth to Mars is the Moon. It has several nice features: it is pretty big, meaning it has a high gravitational field; it is close, at least compared to Mars, so we don't have to spend a lot of fuel to get there; and it has no atmosphere, so the spacecraft can dive close to its surface on the voyage out. This is the basis of the popular Mission to Mars ride at EPCOT in Disneyworld.

Another means of utilizing a third body is by means of a powered slingshot to increase the speed of the spacecraft. In this case a spacecraft makes a flyby of a planet, carrying out a Δv maneuver at the point of closest approach (the periapse) of the orbit. This effectively multiplies the effectiveness of the maneuver. Let us say that the spacecraft is moving at speed v and makes a Δv change in its speed very far from a planet (in the same direction that the spacecraft is already moving). The new speed is $v' = v + \Delta v$. However, if it is traveling at speed v far away from a planet, at periapse it will be moving at speed

$$V = \sqrt{v^2 + v_e^2},$$

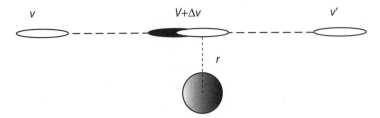

Figure 9.5. Powered slingshot maneuver.

where $v_e = \sqrt{2GM_p/r}$ is the escape velocity from a planet with mass M_p (at distance r from its center). If it makes the same Δv maneuver at periapse, the speed of the spacecraft when it gets far away from the planet will be

$$v' = \sqrt{v^2 + 2\Delta v\sqrt{v^2 + v_e^2} + \Delta v^2},\tag{9.6}$$

which is greater than $v + \Delta v$.[4] The planet can be used to boost the Δv maneuver.

In the late 1980s Edward Belbruno pointed out that, at least in the case of travel from the Earth to the Moon, there were classes of orbits (which he called "chaotic orbits") that included the effects of the Moon and Sun on the shape of the orbits. These orbits could be used to get a spaceship from the Earth to the Moon at a fraction of the energy cost of the Hohmann orbit. The price is a much longer transit time, about a month instead of three days. Such chaotic orbits were used to successfully transfer the Japanese Hiten probe from an orbit around the Earth to the Moon in 1990 [35, pp. 2ff.].

Unfortunately, none of these options to reduce the fuel and energy costs reduces the most critical parameter for manned spaceflight, the time it takes. If we want to send people to the planets on relatively low-energy orbits, we are constrained by the laws of orbital dynamics to trips that take years, and we therefore must deal with the vexing problem of keeping people alive in space for years at a time. As we saw in chapter 4, this comes with a very high sticker price, such that to date, none of the space missions to Mars or farther out has been manned.

9.7   COSTS

Again the question arises, are the benefits worth the costs? Science fiction writers of the 1950s and 1960s blithely wrote about the exploration of Mars as if it were the exploration of some uncharted region on Earth, like the Conquistadors in Central and South America. Let's consider the ground conditions on Mars to see whether this is reasonable.

The average temperature on Mars is about the same as that of Antarctica. The atmosphere is almost too thin to be considered poisonous; at "sea level" on Mars the pressure is about the same as at the top of Mount Everest, and the atmosphere is an unbreathable mix mostly composed of carbon dioxide. There is very little surface water, even in the form of ice, even at the poles (which are mostly frozen carbon dioxide.) There is strong evidence that water once flowed on the surface of Mars billions of years ago, but it's gone now. It is barely possible that life exists on Mars at the bacterial level. We have sent several unmanned probes there to see, and so far no evidence of life has been found.

No one can stay on Mars without protection, meaning warmth and atmosphere. If astronauts were to stay there for many months, they would have two choices: either stay in a craft in orbit around Mars and explore the surface using robots, or descend to the surface and construct a habitable base. We already explore Mars using robots; the only advantage to controlling the robots from Mars orbit is that the delay time is considerably shorter (a second or so, compared to 4.3 minutes when Mars is at its closest to Earth). I'm not sure that this is enough of a gain over exploration of Mars using a robot probe to justify the factor of at least 100 in costs for such a mission.

If they descended to the surface, the question becomes what they could do on the surface that a robot probe couldn't. This is very relevant considering the added costs of creating a base they could live in. In previous chapters we've considered the costs of bringing an Earth-like environment into space; any base would most likely cost billions of dollars, and there would be no a priori way to test whether it would work for the long time which our hardy explorers would need to stay in it.

The whole question of manned space exploration is a vexing one. Most science fiction writers of the 1940s through the 1980s and later foresaw the existence of colonies on Mars and on the moons of the giant planets. Robert Heinlein's novel *Farmer in the Sky*, for example, posits that mankind has terraformed Jupiter's moon Ganymede and used it for farming purposes to feed overpopulated Earth. On a more recent note, TV shows such as the *Star Trek* franchise, *Babylon 5*, and the old and new *Battlestar Galactica* series implicitly assume that mankind will spread out beyond planet Earth. This assumption, which is deeply ingrained in science fiction culture, comes from a time before the "rise of the robots"; no one thought that computing power, the basis of today's unmanned robotic explorers, would increase as rapidly as it has, or be as cheap as it is. It isn't clear there is anything that can be done by a human in space that can't be done more cheaply and easily by a machine. At least one writer has called the notion of manned space travel "old-fashioned". From a scientific standpoint, I think this is correct. Manned space travel is too costly and has too few benefits to justify its existence.

Most readers might now expect a closing paragraph in which I extoll the nonscientific benefits of manned space exploration: the thrill of the exploration of the unknown; the idea that mankind needs new frontiers if it is not to stagnate; the worry that if mankind is stuck on one planet, a disaster could destroy us. These are appealing ideas. But manned space exploration clearly will not happen unless we find better ways of getting off-planet and creating homelike places elsewhere. I'd like to construct an analogy: we are in the same situation with regard to manned spaceflight today as Charles Babbage was with respect to computing in the 1860s. He invented the basic ideas for the modern computer and tried to implement them using the mechanical technology of his day. The technology was marginally not good enough to allow his analytical engine to be built. We seem to be in the same situation today: chemical rockets with exhaust speeds of a few thousand meters per second are marginally good enough to launch unmanned probes traveling slowly through the Solar System but are completely inadequate for manned missions. In the next chapter we examine solutions that have been proposed for getting out there more quickly.

NOTES

1. To quote the song "Shoehorn with Teeth" by the band They Might Be Giants: "Tour the world / In a Heavy Metal band / They run out of gas / The plane can never land." Untrue for planes, true for spacecraft.

2. For a full calculation, including escape velocities from Mars and Earth, see the paper "Journey to Mars" by Arthur Stinner and John Begoray [227]. One issue with this paper is that the authors do not discuss the fuel costs correctly. They assume fuel costs proportional to kinetic energy of the spacecraft instead of using the rocket equation.

3. Technically speaking, this is a *collision problem*, even though the spacecraft doesn't actually collide with the planet (at least, so we hope). The original idea for this may be due to the mathematician Stanislaw Ulam in his Los Alamos report LAMS-2219 [75, pp. 25–26] [240, chap. 9].

4. This seems paradoxical and a violation of the conservation of energy. What is happening is that the spacecraft is losing gravitational potential energy by getting closer to the planet. By burning the fuel closer to the planet you are gaining not only the kinetic energy from the spent fuel but also the potential energy it has "lost" because the spent fuel will remain close to the planet.

CHAPTER TEN

ADVANCED PROPULSION SYSTEMS

10.1 GETTING THERE QUICKLY

In the last chapter we considered the Hohmann transfer orbit as a means of traveling from Earth to Mars or another planet. It was an energy-efficient method of getting there from here, but it had one big disadvantage: it took a long time. This was because the planets had to be in the right relative positions when the rocket was launched, and also because when the rocket was launched it became a planet in effect: a satellite of the Sun, acted on only by the force of gravity (except for short times when changing orbits). A trip to Mars took about 0.7 years, and trips to the outer planets took much longer. Everything is governed by Kepler's third law.

The reason it took so long is that the times when the spacecraft was accelerated by its engines were relatively short. For a minimum-energy, Hohmann-type orbit, the spacecraft is in free-fall orbit around the Sun except when making (brief) Δv maneuvers. However, if we want to travel to Mars over the weekend and be back in time to watch *My Favorite Martian* reruns on Monday night, we need spacecraft that are capable of acceleration for very large stretches of time, meaning they will need to expend a lot of energy. It is pretty easy to see that conventional rockets simply won't work for this particular job.

10.2   WHY CHEMICAL PROPULSION WON'T WORK

Let's take a hypothetical spacecraft that can accelerate continuously at 1 g so that we feel our normal weight along the voyage. The three questions we want to look at are:

1.   How long will it take?
2.   What is the greatest speed we reach along the voyage?
3.   How much fuel will we need for a 10,000 kg payload?

We'll use the final question as a criterion to evaluate the promise of different types of propulsion systems.

The average distance of Mars from the Sun is 1.52 AU, that is, 1.52 times the average distance of Earth from the Sun. Therefore, at closest approach, Mars is 0.52 AU from Earth, or (about) 7.5×10^{10} m, that is, 75 million km, or about 40 million miles. The acceleration of gravity is about 10 m/s^2; I'll assume that the ship accelerates halfway, then flips around and decelerates the other half. We don't want to zoom by Mars at high speed, after all this effort.

We start with a formula from freshman physics:

$$d = \frac{1}{2}at^2.$$
(10.1)

That is, at constant acceleration the distance a body travels (d) is equal to one-half the acceleration (a) multiplied by the voyage time (t), squared. I'm assuming that the spaceship isn't moving when the trip starts. Inverting this,

$$t = \sqrt{\frac{2d}{a}}.$$
(10.2)

One interesting thing about this equation: because the spacecraft is continuously accelerating on the way out (that is, traveling faster and faster all the time), going four times the distance takes only twice the time. From this, the time it takes to go halfway to Mars (remember, we are accelerating halfway and decelerating the other half) is

$$t_{1/2} = \sqrt{\frac{2 \times 3.75 \times 10^{10}\,\text{m}}{10\,\text{m/s}^2}} = 86{,}000\,\text{s} = 1\,\text{day}.$$

The total transit time is $t = 2t_{1/2} = 2$ days. In more useful units, if we wish to go a distance of d AU at an acceleration of $1\,g$, the time it will take is

$$t\,(\text{days}) = 2.8\sqrt{d}\,(\text{AU}).$$

So, *if* we can accelerate at $1\,g$, the Solar System is ours!

The maximum velocity on our trip to Mars is at the midpoint and is given by

$$v_{max} = at_{1/2} = (10\,\text{m/s}^2) \times 86{,}000\,\text{s} = 860{,}000\,\text{m/s}. \tag{10.3}$$

As fast as this is, it is only about 0.2% of the speed of light. To achieve this velocity for a 10,000 kg payload with chemical rockets (i.e., a typical exhaust velocity of about 3,000 m/s), we need a mass ratio of approximately $e^{287} \approx 10^{124}$, which is clearly impossible. This is a physical impossibility, not merely a practical one, as there is not enough mass in the universe to achieve this.

10.3   THE MOST FAMOUS FORMULA IN PHYSICS

Chemical rockets rely on chemical potential energy for propulsion. Chemical energy is the energy released when you rearrange molecules to form other molecules—in other words, the energy released when atoms in some sort of compound change their relative positions with each other. This places an intrinsic limit on the amount of energy that can be released by a chemical reaction, as it is due to the electrical potential energy between different atoms.

This is a limiting factor because atoms, although teensy on a human scale, are pretty far apart in the microworld. We can do a rough estimate of the electrostatic potential energy in a kilogram of matter by thinking about two adjacent atoms. Typically, atoms are separated by about 10^{-9} m in a solid or liquid. Let's approximate this with the idea that the atoms are represented by two charges separated by a distance r about equal to the average interatomic spacing. This will underestimate the energy, but not by a whole lot:

$$E = \frac{ke^2}{r}. \tag{10.4}$$

Here, $k = 9\times10^9$ Jm/Coul^2, $e = 1.6\times10^{-19}$ Coul, and $r = 10^{-9}$ m. This gives about 2×10^{-19} as the approximate potential energy between a pair of atoms. Since there are about 10^{26} atoms per kilogram of a solid, this implies about 2×10^7 J/kg (i.e., 20 MJ/kg) available in the form of chemical energy. This value is pretty good for a crude estimate: gasoline, for example, liberates about 80 MJ/kg when combusted.

However, there is a lot more energy in matter; indeed, matter is a form of energy. This is what Einstein's famous formula,

$$E = Mc^2, \tag{10.5}$$

states: there is an energy equivalent in matter to its mass (M) multiplied by the speed of light squared (c^2). In other words, 1 kg is equivalent to 9×10^{16} J—90 million billion J, or about a billion times the energy present in chemical reactions. If we could somehow liberate even a small fraction of the energy present, we would not only have a working space drive but could also solve all of Earth's energy problems, essentially forever (but more on that later.) But how?

10.4 ADVANCED PROPULSION IDEAS

10.4.1 Nuclear Energy Propulsion Systems

The idea for nuclear propulsion of spacecraft dates back to at least 1945, as Richard Feynman documents in his book *"Surely You're Joking,*

Mr. Feynman!":

> During the war, at Los Alamos, there was a very nice fella in charge of the patent office for the government, named Captain Smith. Smith sent around a notice to everybody that said something like, "We in the patent office would like to patent every idea you have for the United States government, for which you are working now [concerning nuclear energy].... Just come to my office and tell me the idea."

> ...I say to him, "That note you sent around: That's kind of crazy to have us come in and tell you every idea.... There are so many ideas about nuclear energy that are so perfectly obvious, that I'd be here all day telling you stuff."

> "LIKE WHAT?"

> "Nothin' to it!" I say. "Example: nuclear reactor ... under water ... water goes in ... steam goes out the other side ... Pshshshsht—it's a submarine. Or: nuclear reactor ... air comes rushing in the front ... heated up by nuclear reaction ... out the back it goes ... Boom! Through the air—it's an airplane. Or: nuclear reactor ... you have hydrogen go through the thing ... Zoom!— It's a rocket. Or: nuclear reactor ... only instead of using ordinary uranium, you use enriched uranium with beryllium oxide at high temperature to make it more efficient.... It's an electrical power plant. There's a million ideas!" I said, as I went out the door. [84]

Atomic energy is very attractive as an energy source for spacecraft propulsion because of the enormously high specific energy (energy/per kilogram of fuel) compared to chemical fuels. There are two types of nuclear reactions: fission and fusion. In fission, the capture of a neutron (n) causes large, unstable nuclei to fall apart, while in fusion reactions, energy is generated by the fusing together of lighter nuclei into heavier ones.

I'm going to digress for a moment on the structure of the atom and the atomic nucleus. Atoms are mostly empty space. An atom is electrically neutral, but all of the negative charge is on the outside: the electron shells that make chemistry possible extend out to a distance of about 0.1 nm, or 10^{-10} m, from the center of the atom. All of the positive charge, in the form of protons, is at the center of the atom; for each electron in a neutral atom there is one proton as well. The protons are confined to a space that is about 10^{-15} m in radius, or 1/100,000 of the extent of the electron shells. Because like charges repel each other, something must act to "glue" the nucleus together. This glue is the neutron. For every proton, there is at least one neutron in the nucleus that provides a force (called the "strong nuclear force") that keeps the protons from exploding outward. Neutrons are neutral (i.e., they carry no charge) and are about the same mass as the proton, which is about 1,800 times heavier than the electron.

The total electrostatic potential energy of the nucleus is very much larger than the electrostatic potential energy between a pair of atoms— about 100,000 to 1,000,000 times bigger because of the size factor involved. The energy liberated from the nucleus is still pretty small compared to the ultimate amount of energy stored in matter, but is large compared to the chemical energy.

Currently, all commercial nuclear reactors work through the fission of heavier elements into lighter ones. As elements become heavier and heavier (i.e., having more and more protons and neutrons in the nucleus), they become unstable and the largest ones can fall apart spontaneously. This process liberates energy: by fissioning into two nuclei, one large nucleus moves about half the protons far away from the other half.

10.4.2 Fission Reactions

In each process, the energy released can be calculated by the difference in mass between the end products and the initial nuclei. A typical fission reaction used in reactors is [246, pp. 1198–1199]:

$$n + {}^{235}U \rightarrow {}^{140}Ce + {}^{94}Zr + 2n + 208\,\text{MeV} \tag{10.6}$$

(keep in mind that $1\,\text{MeV} = 1.6 \times 10^{-13}\,\text{J}$).

Because an extra neutron is generated by the fission process, the process can be self-sustaining if enough ^{235}U is present; this amount is called a *critical mass*. Nuclear reactors use an amount very slightly over criticality to generate energy in a controlled way, whereas nuclear bombs suddenly throw together two or more very slightly subcritical masses so that the chain reaction is fast, resulting in a huge, short burst of energy. Both types of reactions have been proposed for propulsion systems.

The energy density of the fission of 1 kg of ^{235}U can be calculated from the molar mass (235 g/mol) of this isotope:

$$\varepsilon = 8.5 \times 10^{13} \, \text{J/kg}.$$

However, in a reactor, only about 3.5% of the uranium is enriched into ^{235}U; most is in the form of ^{238}U, which is less reactive, meaning that the energy density drops to about 3×10^{12} J/kg. This is still about 10^5 times greater than the specific energy available from chemical reactions.

Most work on nuclear propulsion systems for spacecraft are in drives that use a nuclear reactor to heat hydrogen gas as an exhaust fuel. Hydrogen is used because, as the smallest atom, it achieves the highest exhaust velocity for a given amount of energy dumped into it. For reasons discussed below, the exhaust velocity is limited to about 8,500 m/s, or roughly twice the maximum achievable by chemical fuels.

Work on nuclear spacecraft engines was stopped by the Nuclear Test Ban treaties of the 1970s. Up until 1972, NASA was working on a series of nuclear rocket engines dubbed the NERVA, or Nuclear Engine for Rocket Vehicle Applications series. Interest in them has revived because of a series of initiatives started during the second Bush administration and carried on by the Obama administration to have manned missions to the Moon and Mars. Nuclear rockets could potentially shorten the time to Mars to under 100 days by allowing a Δv for the initial orbital maneuver greater than 34 km/s.

Both NERVA and the Orion concept have several advantages over chemical propellants. One is that they tend to work better for larger payloads (meaning they can be designed to give higher thrust and impulse). There is a minimum size at which nuclear reactors can be built, but scaling them up is comparably less difficult. There are, of course, obvious disadvantages of using nuclear fuel, but they tend to be overstated.

10.4.3 NERVA

Robert Heinlein's novel *Rocket Ship Galileo* is about a group of Boy Scouts and their nuclear physicist mentor who build a spacecraft and go to the Moon, defeat a group of Nazi astronauts while there, and return triumphantly home. It is the first science fiction novel I know of to use the idea of a nuclear reactor to heat up and eject propellant, even though this predated the NERVA program by 20-odd years [109]. However, this idea hasn't caught on as much as other propulsion ideas in the science fiction literature, principally because of the limitations of conventional nuclear reactors.

The NERVA program was initiated by NASA in the 1960s to build a nuclear-powered spacecraft for a planned manned mission to Mars (scheduled to take place around 1970). A small nuclear reactor was designed as a power source for the spacecraft; the reactor heated liquid hydrogen to a temperature of 2,200 K and expelled it through a nozzle to generate thrust. The initial design showed some promise: the exhaust velocity was $u \approx 8{,}600$ m/s, which is nearly twice the best value obtainable from rocket fuels. It also had a relatively high thrust of 73 kN. The NERVA propulsion system was also proposed as a potential engine for the Space Shuttle but was killed by post-detente cuts in NASA's budget and a general distrust of nuclear power in the 1970s.

The biggest issue with this type of propulsion system is that although a lot of energy is liberated in nuclear processes, it isn't obvious how to use it. The limitations imposed on fission reactor spacecraft have more to do with materials science than with energy usage: for one thing, the energy liberation rate is limited by the melting point of the material one makes the reactor from (ultimately, the uranium alloy used as fuel). Also, the neutrons that are produced embrittle the spacecraft engine, which places limits on the energy generation rate. Finally, the method simply heats the reaction mass (the hydrogen fuel), which may not be the best way to use all of this energy. There should be a better way, a more clever design, that uses all the energy directly. NERVA represents an incremental advantage over chemical rockets because it is a relatively conservative design. The design of the propulsion system for the Orion project, however, represents a real departure from most rocket concepts.

10.5   OLD "BANG-BANG": THE ORION DRIVE

The most interesting nuclear propulsion system was invented by the mathematician Stanislaw Ulam and C. J. Everett and developed by physicists Freeman Dyson and Ted Taylor in the 1950s [41] [75, pp. 22–24] [240, chap. 7] It is no longer taken seriously by anyone except science fiction writers; however, it does represent thinking big. The idea was to build a spacecraft with a big, highly shielded plate at the back and *blow up a nuclear bomb behind it* to push the ship forward. I swear to God I am not making this up. This was referred to euphemistically by NASA as a "nuclear pulse drive" [210]. Dyson, currently a fellow at the Institute for Advanced Study at Princeton University, thought it could be done safely, and ran a pilot program to study it. The full story is related in the books *The Curve of Binding Energy, The Canoe and the Starship*, and *Project Orion* [75][164]. Dyson is a firm believer in thinking big in science; we will examine the idea of a "Dyson sphere" (another science fiction fave) later on in this book. His idea to make Orion work was to use a sequence of small nuclear bombs to give a more or less uniform acceleration to the ship. One thing to note here: I realize that at this point I've discussed only nuclear fission (the power source of atomic bombs), whereas Orion would be using specially made hydrogen bombs.

This idea has proved incredibly popular in the science fiction community. I've read two novels that use the idea, S.M. Stirling's *The Stone Dogs* [228] and Larry Niven and Jerry Pournelle's *Footfall* [187]. *The Stone Dogs* is set in an alternate–history present/future in which the world is divided between the United States and its allies and the Domination of the Draka, a highly advanced technological slave-owning society based in South Africa. Stirling assumes that because of the rivalry between the two, certain forms of military technology (especially developments in space) are accelerated relative to our own world. By the 1960s, nuclear pulse drives have been developed and are later used to put colonies on the Moon and settle the asteroid belt.

In one of their patented cast-of-thousands novels, Larry Niven and Jerry Pournelle in *Footfall* present an Earth being invaded by aliens resembling miniature elephants. The aliens use a Bussard ramjet for propulsion, which I discuss in the next chapter. The humans in the story, led by a team of science fiction writers and fans, institute a crash program to build an Orion-type spacecraft to combat the aliens. There are many others examples of the genre. Science fiction writers (and readers) like space travel and things that go bang; combining them is almost irresistable.

The energy released by a Hiroshima-sized nuclear bomb is about 10^{13} J. If this could be converted to kinetic energy, it would get our canonical 10,000 kg payload moving at a speed of about 50 km/s. Unfortunately, the energy is liberated in a few microseconds, meaning that everyone would be crushed to jelly from the high acceleration. Also, a payload as small as 10,000 kg is unrealistic for a manned vehicle. In reality, the initial studies were of payloads ranging in mass from 10 tons to about 400 tons, with mass ratios of about 10. They relied on small nuclear devices blown up at a rate of roughly one per second and an effective exhaust velocity of about 20,000 m/s, or about seven times what we've assumed for chemical propellants [9]. It's pretty clear that if such a system could be made to work, then one could send fast rockets to the far corners of the Solar System, if not to the stars. You would need to launch the rocket from orbit, as the radioactive fallout is not something you want in your back yard. Three major technical issues have to be overcome in designing a spacecraft using nuclear bombs: shielding the passengers from the radiation produced in blowing up a nuclear bomb every second or so behind them; keeping the average thrust low enough so that people wouldn't be smashed to jelly by the high acceleration; and designing a system to throw a nuclear bomb behind the ship once a second and detonate it there—a difficult design problem. According to George Dyson, the Orion team consulted the Coca-Cola Company on the bomb delivery system for the spacecraft [75, pp. 177–178]. A final problem is that getting Congress (or any agency) to fund such a program is improbable, to say the very least. This is what eventually killed the program. It is fun to think about, though; a lot of sweet design problems stem from trying to figure out how to launch spacecraft with nuclear bombs. A few are listed on the book website.

The preliminary studies for Orion were for ships traveling to the Moon or on conventional Hohmann orbits to Mars; the pulse drives were to be turned on for only a few minutes during the two Δv boosts, and the total trip time was 450 days. However, Dyson and Taylor were ambitious: they wanted ships that could accelerate continuously, and it is clear that this is what they expected the Orion project to eventually deliver.

In overall design, the main part of the ship rests on top of a "pusher plate" that serves both as radiation shielding and as something for the bomb to push against. A bomb is dropped behind the plate and blown up. The pusher is attached to the rest of the ship by an arrangement of springs and mass dampers that smooths out the inherently jerky nature of the acceleration.

The overall (theoretical) performance is very impressive. The specific impulse is very high ($I_{sp} \approx 10^3$ s, or $u \approx 10^4$ m/s or higher). This means that for typical Δv corrections of 1–10 km/s, the ratio of fuel mass to payload mass is of order one. As one might expect, thrust is also high.

One thing unique to Orion is that there is no way to make a small spacecraft using this concept because of the difficulty of making low-yield nuclear bombs. The first design (for a Mars mission) called for a 4,000-ton (4×10^6 kg) spacecraft using 0.1 kT yield bombs specifically designed for the project. By comparison, the lunar landers for the Apollo missions massed about 14,000 kg. Several *thousand* of these bombs would be detonated during the round-trip mission. Because of the high impulse, one can also use faster orbits than the minimum-energy Hohmann transfer.

The development of Orion came to a halt in the early 1960s when priority was given to the Apollo program, which used more standard chemical propulsion systems to reach the Moon. However, Orion is the *only* feasible high-impulse, high-thrust propulsion system studied in detail to date.

10.6   PROSPECTS FOR INTERPLANETARY TRAVEL

There are definite advantages to using nuclear propulsion; the effective exhaust velocities range from twice to about five to ten times what one can get by using traditional chemical propellants. They really are the only way one could envision large-scale manned interplanetary travel, in that they could cut the travel time between Earth and the other planets in the Solar System by about an order of magnitude. If one could use Orion to accelerate a spacecraft at 1 g indefinitely, one could cut the time by a factor of 100 or so.

However, the nearest stars are thousands of times more distant than the farthest planets in the Solar System. To get to the stars we need even more energy than even Orion can provide, but for this saga, we take up the story again in the next chapter.

CHAPTER ELEVEN

SPECULATIVE PROPULSION SYSTEMS

As far as we know, nothing travels faster than light. Aside from energy issues, it takes a long time to get anywhere interesting. The nearest star apart from the Sun is the triple-star system Alpha Centauri, located 4.3 light-years away. There are about 32 star systems within 15 light-years of the Sun, and roughly 600 within 100 light-years. This means it will take years to get to the nearest stars even traveling at speeds close to light. There is one saving grace: because of relativity, the trip won't seem as long to the voyagers. However, to get to these speeds, even the Orion drive is insufficient.

11.1 MORE SPECULATIVE PROPULSION SYSTEMS

11.1.1 Fusion Reactors

Fusion reactions generate energy from building up heavier nuclei from lighter ones. Again, there is net energy produced in the reaction because the reaction products are lighter than the reactants. This is the energy source inside the Sun. The reaction there is the proton-proton cycle,

$$4^1 H \rightarrow {}^4He + 2e^+ 2\nu + 6\gamma \ (26.71 \, \text{MeV}). \tag{11.1}$$

This reaction needs high densities and is relatively slow, making it unusable for Earth-based fusion applications. Most researchers concentrate on the deuterium-deuterium reaction:

$$^2H + {}^2H \rightarrow {}^3He + n \ (3.27 \, \text{MeV}), \tag{11.2}$$

or the deuterium-tritium reaction:

$$^2H + {}^3H \rightarrow {}^4He + n \ (17.59 \, \text{MeV}). \tag{11.3}$$

The latter process has (in principle) an energy density of 3×10^{14} J/kg of total fuel, or one to two orders of magnitude greater than fission processes [246]. If you used the reaction products as ejection mass for the spacecraft, you would have an exhaust velocity of about 2×10^7 m/s, or 20,000 km/s, roughly 7% of the speed of light. In principle, half a kilogram of this fuel could send a 1,000 kg spacecraft to the Moon. The proton-proton cycle has a higher exhaust velocity of approximately 12% of the speed of light. One tricky part is that while deuterium is pretty common, tritium is not: it is an unstable isotope with a half-life of 12 years. Even given this issue, a 1974 study of how to build a spacecraft capable of interstellar flight, Project Daedalus, fixed on a deuterium-tritium fusion reactor as the only feasible means of powering the spacecraft.

Fusion energy can only be generated at temperatures of millions of degrees and high densities, which is why it is difficult to generate it controllably; however, if one could harness it, it would supply an essentially limitless source of power. This is why the development of controllable fusion has been something of a holy grail for physicists for the past fifty years; commercial fusion power plants would essentially solve the world's energy problems. However, it is a holy grail in many senses: no one really has any idea how to do it, although much work has been done on the problem. As the joke goes, fusion power is always 20 years away. Needless to say, no one knows how to build a fusion-powered rocket either, although there have been suggestions on that issue as well.

So fusion rockets are a good way to go, *if* we had any idea how to make them. The first study I know of on making an interstellar probe using a fusion engine was Project Daedalus in 1975. This was a serious study undertaken by a number of scientists and engineers to put together a "proof-of-principle" design for a probe capable of a flyby mission to Barnard's star, six light-years from Earth. The speed chosen was 15% of light-speed, making it an approximately 40-year journey. There are definitely some science-fictiony aspects of the project; for example, it called for mining Jupiter for tritium for the fuel supply. This seems to be on the borderline of possible, but exactly which side is anyone's guess.

However, the implication of an interstellar probe like this one is that we possess an extremely energy-rich society. The cost of Project Daedalus was estimated at $10 trillion. Using the rule of thumb that prices for everything double every 20 years, the estimate comes in at about $40 trillion today, dwarfing the U.S. GDP. This amount of money is about equal to the GDP of the entire world. Energetics tell us why this is so: the total energy contained in the payload is about 10% of the total world energy usage for one year. This is too expensive for any current world civilization to undertake, and it may well be too expensive for any civilization to undertake under any circumstances.

11.1.2 The Bussard Ramjet

The nearest star apart from our sun, Alpha Centauri, is about 4.3 light-years away, At an average speed of 10% of the speed of light, it would take 43 years for a spacecraft to get there and the same to return, if we want it to return. This speed is possible for a fusion-powered craft; assuming the exhaust speed is as stated above, 2×10^7 m/s, the mass ratio isn't too prohibitive. The spacecraft needs about 3.5 kg of fuel for every kilogram of payload if we don't decelerate. If we do decelerate, then we need this quantity squared, or about 12 for a fuel/payload ratio; if we want it to return, then the quantity is raised to the fourth power (about 144). However, the round trip is clearly longer than any human life, and even relativistic time dilation will not help us much. Unfortunately, going much faster gets very difficult. At this point, for speeds near the speed of light, our original formula for the rocket equation doesn't work any more, so we need to use a version that is corrected to take the special theory of relativity into account. This was first derived by Ackeret in a 1946 paper; you can find a derivation of the same formula in English in the paper "Relativistic Rocket Theory" by Bade in *The American Journal of Physics* [17][29]:

$$\frac{v}{c} = \frac{(m_i/m_f)^{2u/c} - 1}{(m_i/m_f)^{2u/c} + 1}.$$ (11.4)

The variables are:

- u: exhaust velocity (assumed to be 2×10^7 m/s);
- v: final velocity reached by the spacecraft;
- m_i: initial (payload + fuel) mass of the spacecraft;
- m_f: final (payload) mass of the spacecraft;
- c: speed of light (3×10^8 m/s).

I'll going to define three new variables, to make life easy on us:

- R: mass ratio m_i/m_f;
- α: ratio of exhaust speed to speed of light ($= u/c$);
- β: ratio of final speed to speed of light ($= v/c$).

We can then solve for the mass ratio needed to get to any fraction of the speed of light:

$$R = \left(\frac{1 + \beta}{1 - \beta}\right)^{1/2\alpha}.$$ (11.5)

The issue is that nothing can go faster than the speed of light. The non-relativistic rocket equation doesn't take this into account; basically, as you go faster and faster, the mass ratio increases more and more, to the point that going at the speed of light would require an infinite mass ratio. Simply to get to 90% of the speed of light with the exhaust velocity given above, we would need a mass ratio of 3.9 billion.

This clearly won't work. So what are we to do? In 1960 the physicist Robert Bussard had an ingenious idea: "empty" space isn't really empty. There is very thinly spread matter in interstellar space, mostly hydrogen. On average, there is about one atom per cubic centimeter in interstellar space (or about $10^6/m^3$), although in dense molecular clouds, there can be as much as $10^9/m^3$, or even much more [130, pp. 435–438]. So: en route, scoop up the material between the stars and use it as fuel [44].

Bussard envisioned a large "scoop" or funnel of some kind extending for hundreds or even thousands of kilometers ahead of the ship, gathering and compressing the hydrogen to the point that it underwent fusion, serving as fuel for the spacecraft. Figure 16 shows a schematic of a Bussard ramjet. This idea has proved incredibly popular in the science fiction literature as it seems to be the only plausible way, apart from matter-antimatter reactions, to get a spacecraft to travel at an appreciable fraction of the speed of light. The idea has been used by a large number of science fiction writers (including Poul Anderson in his novel *Tau Zero*), but it was certainly used most extensively and popularized enormously by the writer Larry Niven in his *Known Space* stories. Indeed, it is almost easier to list his stories that don't use this concept. A sampling:

- In the novel *Protector*, Phssthpok the Pak uses a ramjet to get from the center of the galaxy to Earth (over the course of 30,000 years). Later in the novel, the Brennan-monster and the Pak scouts use ramjets to fight their extended interstellar skirmishes [179].
- In *A Gift From Earth* (and other novels and stories), it is mentioned that unmanned Bussard ramjets explore interstellar space for inhabitable planets (or, as mentioned in the novel, "inhabitable points"), which are later settled by "slowboats" carrying human settlers.
- In the story "The Ethics of Madness," a paranoid steals a spacecraft and kills a friend's family using it, and is then literally chased to the ends of the universe by the other in a second ramjet.
- In the non-*Known Space* novel *A World out of Time*, the world government known as "The State" uses revived "corpsicles" to pilot ramjets to seed potentially Earth-like planets with bacterial life in the hopes of terraforming them for settlement.[1]

And this is just a sampling.

I'm going to present a nonrelativistic analysis of the Bussard ramjet: for a fully relativistic one, see Bussard's original paper or any of the references below.

At first glance, because the ramjet picks up its fuel en route, it appears that we can keep accelerating forever, not subject to the limitations imposed by the rocket equation. Let's make some assumptions:

1. The ramjet, of total mass M, is under way and traveling at some speed v relative to the Earth.
2. It has a funnel of some sort that scoops matter into the fusion engines. The funnel has area A.
3. Fusion uses a fraction f of the hydrogen for fuel, which yields energy with an efficiency α and ejects the rest to the rear at speed u (which depends on both f and α).
4. The interstellar hydrogen has a mass density ρ.

The thrust of the spacecraft is proportional to two things, the fuel speed, u, and the rate at which mass is being fed to the engine, dm/dt, which can be shown to be equal to ρAv. This makes sense: the faster a ship moves forward, the more hydrogen it scoops up; the large the funnel area, the more hydrogen it scoops up; and the higher the density, the more it scoops up:

$$T = u\frac{dm}{dt} = \rho A(1 - f)uv.$$

The thrust increases with increasing speed, all other things being equal [44]. If we want the acceleration to be constant (meaning that the thrust is constant), we need to control the fraction f of the fuel undergoing fusion. If we intake a certain amount of hydrogen equal to m into the ramjet, the exhaust speed of the fuel can be found (in the nonrelativistic limit):

$$\frac{1}{2}(1 - f)mu^2 = f\alpha mc^2 \tag{11.6}$$

or

$$u = c\sqrt{\frac{2f\alpha}{1-f}}.$$

This means that we can write the equation for the acceleration $(=T/M)$ in the form

$$a = \frac{A\rho}{M}\sqrt{2\alpha f(1-f)vc}. \tag{11.7}$$

If we want the acceleration to be constant (say, $1\,g$ for the entire flight) then we must make f, the fraction of hydrogen used as fuel used by the ramjet, depend on v. To calculate f we must solve the quadratic equation

$$f(1-f) = \frac{1}{2\alpha}\left(\frac{aM}{\rho Avc}\right)^2. \tag{11.8}$$

If we want a certain value of acceleration, there will be a minimum speed below which this is impossible to do. The reason is simple: at low speeds, the amount of mass going into the ramscoop will not provide enough thrust.

There is another force acting on the ship. By analogy with motion through fluids on Earth, there will be a "drag force" exerted on the ship. This drag force will have the general form

$$D = \beta\rho Av^2, \tag{11.9}$$

where β is a dimensionless coefficient of order 1; its exact value will depend on the design of the ramscoop. This implies a maximum velocity for the ship: as the speed gets higher, the drag force increases until it exactly balances the thrust. At this point,

$$\beta v^2 = \sqrt{2\alpha f(1-f)vc}. \tag{11.10}$$

The maximum value of $f(1 - f)$ occurs at $f = 1/2$, so the maximum speed of the ship is

$$v_{max} = \frac{\sqrt{\alpha/2}}{\beta} \times c. \qquad (11.11)$$

At best, $\alpha = 7 \times 10^{-3}$ for the proton-proton cycle, so the maximum speed is going to be around 6% of the speed of light. This depends on β, of course, so there might be some way of "streamlining" the ramscoop to minimize this problem.

There are a number of problems with the idea, which probably make it untenable, apart from the speed limitation mentioned above:

- Unknown in 1960, when Bussard first published his paper, is that the Solar System is in the middle of a high-temperature "bubble" of interstellar gas about 650 light-years across whose density is relatively low, about .006 molecules per cm^3, or about 6,000 $/m^3$ [130, p. 435]. This density is about two orders of magnitude lower than Bussard's estimates [44].
- Most of the matter in interstellar space is hydrogen; however, as noted above, fusing hydrogen is difficult because of the slowness of the reaction.
- Fusion of deuterium is easier, but deuterium is many thousands of times less common in interstellar space than hydrogen.
- Because the ramjet will work only if the ratio of the funnel area to the mass is very large, most papers concentrate on using magnetic fields as funnels. These work only with charged particles; some have suggested using either lasers to ionize the interstellar medium or using strong magnetic fields. Materials limitations on our ability to produce large magnetic fields reduce the acceleration of the ramjet significantly once it reaches relativistic speeds, even if we ignore drag [158]. The scoop diameter also needs to be about 10^7 km at low speeds for it to work [159].
- Losses due to radiation from the conversion of matter into energy will limit the ship to speeds significantly less than the speed of light [215].
- When charged particles are accelerated they lose energy. These *Bremsstrahlung* losses may exceed energy gained by fusion of the interstellar medium by a factor of 10^9 [123].[2]
- Finally, a point I don't think anyone else has ever brought up: an influx of charged particles at high speeds entering the ramscoop magnetic field will

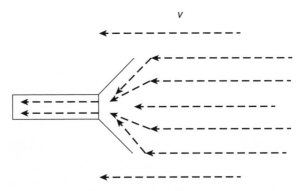

Figure 11.1. Particles funneled into the ramscoop.

distort its shape in the same way that the solar wind distorts the Earth's and Jupiter's magnetic fields [130, p. 275]. It will compress the field in the direction of the ship's motion and drag it behind the ship, making it less effective as a funnel.

More problems with the idea are discussed in detail in John Mauldin's book, *Prospects for Interstellar Travel* [160, pp. 110–116, 326]. If the ramship idea can be made to work, the issues associated with it will probably limit its speed to a small fraction, probably less than 10%, of the speed of light.

Most science fiction writers ignore this and assume that the ramjet can accelerate indefinitely. However, this speed limit puts the kibosh on novels like *Tau Zero* and *A World out of Time*. While this speed is pretty good, if we really want to get to the stars within a human lifetime, we need to go faster—*much* faster. How do we do that?

11.1.3 Matter-Antimatter Drives

Popular culture once again comes to our aid. As everyone knows, the Starship *Enterprise*, serial number NCC-1701 (or 1701A, or 1701D, etc., depending on which movie or show you watch), is powered by the reaction of matter and antimatter. It is in fact unclear from the show whether the ship is powered by these reactions or whether the ship is propelled by them when using its impulse engines. (I'm typing this in the realization that some diehard fan is going to email me, quoting chapter and verse, exactly how it is used.) Maybe we can use antimatter for spacecraft propulsion, just as the Federation of Planets does.

What is antimatter, anyway? It's a long story. Back in the 1930s, one of the major challenges facing physics was how to marry the two newly discovered theories of special relativity and quantum mechanics. An English physicist, Paul Dirac, proposed an equation that described the relativistic quantum behavior of the electron. When the equation was solved it predicted there should be a companion particle to the electron, one with identical mass and other properties but with opposite electrical charge. Because the electron has negative charge, this new particle was dubbed the positron. The Caltech physicist Carl David Anderson detected the positron in cloud chamber experiments in 1932. It quickly became apparent to physicists that all elementary particles must have antiparticles associated with them, and the antiproton and antineutron were discovered within the next few decades.

There's a tendency for nonphysicists to somehow think that antimatter is exotic, or a theory that hasn't been proved yet. All these notions are untrue: antimatter particles by the trillions are produced in accelerator experiments all over the world. This is not to say that much has been produced: in 2001 CERN, a European consortium running the largest particle accelerators in the world, estimated that antimatter production is less than one-billionth of a gram per year at an equivalent cost of several trillion dollars per gram [2].

Why so expensive? Because antimatter doesn't exist in nature, at least to any appreciable amount. Most matter is "ordinary"; antiparticles are produced naturally only in radioactive decays or in processes that have temperatures equivalent to the center of the Sun or higher. It all hinges on $E = Mc^2$. To produce a kilogram of antimatter, you have to supply the energy equivalent to it, or roughly 9×10^{16} J. Put differently all of the energy used by the United States in the course of one year (1.4×10^{20} J) is equivalent to the energy produced by the annihilation of about 750 kg of matter with 750 kg of antimatter. If antimatter existed in large quantities, it could be used to fuel civilization; because it doesn't, the tables are turned. It takes enormous amounts of energy to produce it.

This doesn't mean that it has to be as exquisitely expensive to produce as it currently is. If the energy costs were the only issue involved in producing it, the cost would be less than one-millionth of the current costs: gasoline, for example, has an energy density of about 10^8 J per gallon, meaning that we'd have to burn 900 million gallons of gas to give us the energy equivalent to 1 kg of antimatter. This would cost about $2.25 billion at 2009 U.S. prices of $2.50 per gallon. There have been several serious proposals for large antimatter creation facilities, mostly authored by the late Robert L. Forward, which would bring the cost of antimatter production low enough to make it a viable starship fuel [87].

On to the question of how to use antimatter in propulsion systems. One issue is that we have to get rid of the notion that the "annihilation" of a particle with its antiparticle somehow produces "pure" energy. Such a statement is meaningless. What really happens is that the interaction of a particle with its antiparticle produces different reactant products, similar to what happens in a chemical reaction. The amount of energy available for use in a propulsion system is determined by what the reactants are. The simplest thing to consider is an electron-positron reaction occurring in which the two particles are at rest with respect to each other. The reaction here is

$$e^- + e^+ \rightarrow 2\gamma \ (1.02 \, \text{MeV}).$$

This is shorthand notation for the reaction of an electron and a positron to create 2 gamma rays, that is, two very high-energy photons. If the electron and positron are at rest, the total energy of the gammas will be equal to the equivalent rest-mass energy of the electron and positron, that is, about 1 MeV or 1.6×10^{-13} J. "MeV" is the abbreviation for the unit mega-electron volt, the electron volt being a unit of energy equal to 1.6×10^{-19} J. Most particle masses are quoted in units of MeV or GeV $(= 1,000 \, \text{MeV})$ for the sake of convenience. Physicists learn to quickly convert from these units into more standard units of energy when needed. Because the electron is one of the lightest particles that have mass (only the neutrinos have less mass), an electron-antielectron annihilating at rest can produce only photons. This is good, because in principle, 100% of the rest mass of the electrons can be used as energy for propulsion.

One method of driving the spacecraft forward is to use the momentum of the photons created to propel it. A light beam shining on your hand carries a force, though a very weak one under normal circumstances. If we shine a light beam on a mirror, there will be a force acting to push the mirror forward equal to

$$F = 2P/c, \tag{11.12}$$

where P is the power of the light beam (i.e., the energy per unit time hitting the mirror) and c is the speed of light. Under most circumstances, this force is weak: for example, if we shine a 1 W flashlight on the mirror, the net force is only 3×10^{-9} N. However, if we took 1 kg of matter and reacted it with 1 kg of antimatter to produce a beam of gamma rays directed against the mirror, and did it in a time of 1 second, the net force would be 6×10^8 N. Even if the mirror had a mass of 10,000 kg (10 tons), it would end up moving at a speed of 60,000 m/s!

11.2 MASS RATIOS FOR MATTER-ANTIMATTER PROPULSION SYSTEMS

Proponents of such a technology, including Forward and Robert Frisbee at the Jet Propulsion Laboratory (JPL), have put together elaborate studies of propulsion systems based on matter-antimatter reactions. There are a lot of nitty-gritty details that we won't go into here; in particular, Frisbee's studies seem to indicate that proton-antiproton reactions are better than electron-positron reactions for spaceship drives, even though less of the initial "mass-energy" of the reactants is available for propulsion [88]. For the sake of simplicity I will consider only cases in which the reactants are transformed into high-energy gamma rays and bounced against some sort of mirror to propel the spacecraft.

Using equation (11.5) with $u = c$, we find

$$v = \left(\frac{R^2 - 1}{R^2 + 1} \right) c, \tag{11.13}$$

where c is the speed of light, R is the mass ratio (m_i/m_f), and v is the final speed reached by the spaceship (assuming that it starts from rest). For $v \ll c$, this is approximately $v/c \approx R - 1$. Since $c = 3 \times 10^5$ km/s,

Table 11.1
Final Velocity of an Antimatter Rocket as a Function of Mass Ratio

Ratio (R)	v/c	v (m/s)
1.01	0.01	3.00×10^6
1.05	0.05	1.50×10^7
1.1	0.1	3.00×10^7
1.2	0.18	5.40×10^7
1.3	0.26	7.80×10^7
1.5	0.38	1.14×10^8
2	0.6	1.80×10^8
5	0.92	2.76×10^8
10	0.98	2.94×10^8
20	0.995	About 3×10^8
100	0.9998	About 3×10^8

to get a 1,000 kg payload to Earth's escape speed of 11 km/s takes a mass of 18 grams of normal matter reacted with 18 grams of antimatter.

Table 11.1 shows the real advantages of using a matter-antimatter propulsion system. A mass ratio of 10 gets you to 98% of the speed of light. Unfortunately, this means that for every kilogram of payload mass you need 5 kg of antimatter to react with 5 kg of normal matter, but 5 kg of antimatter exceeds the world's supply by a factor of about a trillion.

There are no naturally occurring sources of antimatter handy. As far as we can tell, the universe is overwhelmingly made of normal matter. This means that any antimatter will have to be created. This is possible; antimatter is created in particle accelerators all over the world, but in very small quantities. Typical generation rates for antimatter are of the order 10^{10}–10^{12} particles per second, or (assuming the particles are antiprotons) 10^{-17}–10^{-15} kg/s. Generously, generating 1 kg of antimatter using current accelerators will take about 10^{15} s, or 30,000,000 years.

However, we're not trying to produce antimatter for spacecraft propulsion right now. Without going into details, essentially the question is one of energy. If we have access to enough energy, we can create antimatter. The questions are how much can we create, and how much will it cost?

11.2.1 The Cost and Time of Producing Antimatter

From the relation $E = Mc^2$, we can't generate more mass than E/c^2 given energy E. So let's make the following assumptions:

1. We have a power plant that supplies energy at a rate of P J/s to supply us with energy.
2. The energy costs us s cents per kilowatt-hour ($= 3.6 \times 10^6$ J).
3. We can convert the energy into mass with efficiency $f < 1$.
4. We want to generate 1 kg of antimatter (which is equivalent to 9×10^{16} J).

The twin questions we want to ask are: How long will it take as a function of P and f. How much will it cost?

In time t, the power plant will generate a total energy Pt joules, or a mass of Pft/c^2 kg of antimatter. Therefore, the time to generate 1 kg is

$$t = 9 \times 10^{16}/Pf \text{ (s)} = 2.85 \times 10^9/Pf \text{ (yr)},$$

using the fact that a year is 3.16×10^7 s. Figure 11.2 shows the time it takes given power plants with energy generation rates running from 10^7 W to 10^{17} W. Two points to note: 10^{13} W is about the total power the world currently produces, and 10^{17} W is about the total power the Earth receives from the Sun.

The total cost (in dollars) can be worked out as well in terms of the energy cost s:

$$M = 2.5 \times 10^8 s/f \text{ (\$)}, \tag{11.14}$$

where M stands for "money." Figure 21.8 shows the total cost for three different values of s: 10^{-3} cents/kW-hr, 10^{-1}, and 10 (which is about the current cost of electricity production in the United States). It should be noted that both figures are on logarithmic scales, as I want to show this as a function of widely differing values of efficiencies, costs, and total power allocated to the task. Also, $f \leq 0.5$ because for every particle of antimatter created, one particle of normal matter is created because of the conservation laws of physics.

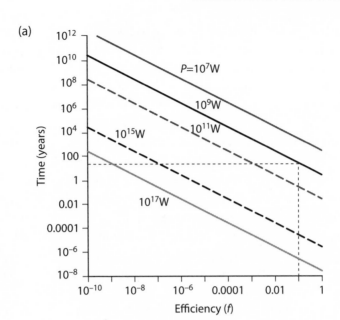

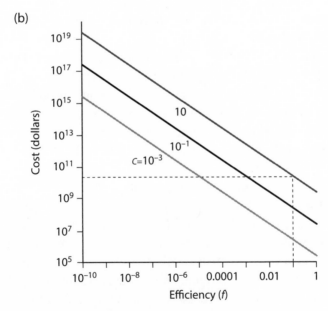

Figure 11.2. Time and cost to produce 1 kg of antimatter.

What are realistic values for f given today's technology? Robert Forward looked into this problem in detail. In a 1984 Air Force technical report he estimated that by using specially designed linear accelerators, one might achieve an overall energy efficiency of $f = 2.5 \times 10^{-4}$ [87, p. 3].[3] To the best of my knowledge, such specially dedicated linear accelerators have never been built. However, a recent paper in *Physical Review Letters* demonstrated anti-electron production using high-energy-density laser pulses with an overall efficiency of 10^{11} electrons per kilojoule, which corresponds to an efficiency of $f = 8 \times 10^{-6}$ [104]. This was a tabletop experiment, so it's not that difficult to imagine that if it were properly scaled up, one could gain a few orders of magnitude in efficiency, to maybe $f = 10^{-4}$. I don't believe any specific physics limits overall efficiency, merely engineering details, but I could be very wrong. This is ignoring the other vexing issue of how we contain all of the antimatter we produce.

The figures make it clear: under almost any reasonable conditions, this is an expensive and lengthy task. I've marked a point on each using dashed lines: assuming that we put all of the power of a 1 GW power plant toward this task (which is just barely possible today) and figured out a way to do this at 10% efficiency (i.e., $f = 0.1$) (which is probably impossible), it would take about 30 years and cost \$25 billion. Using a more reasonable efficiency of 10^{-4} we could produce 1 gram of antimatter in the same time at the same cost, which is more or less the goal Forward was hoping to achieve in his report. Forward's estimates of costs are similar to mine: he calculated that with $f = 10^{-4}$, the cost would be 10^7 \$/mg [87, p. 150].

The only novel I have read that deals with this issue honestly is *Building Harlequin's Moon* by Larry Niven and Brenda Cooper [185]. In this book the starship *John Glenn* is fleeing an Earth ruined by nanotechnology. The ship is powered using a matter-antimatter drive, but it has a problem with it and so must stop en route in another star system to generate enough antimatter to be able to proceed onward. The process of generating the antimatter takes literally hundreds of years. Antimatter generation is power-limited: the rate at which you can create it is limited by the available energy you have on hand. Under most circumstances, this will be pretty slow.

11.3 RADIATION PROBLEMS

Edward Purcell, the Nobel laureate physicist, pointed out in 1963 that even apart from expenses, there are some real problems with trying to get anywhere really fast [46, pp. 121–143]. Those problems have to do with exposure to radiation, both to our astronauts and (if we're not careful) to the entire Earth as well.

Let's say we have collected our antimatter and we want to send a group of explorers out to explore the cosmos. Let's say we want them to travel there and back at 99.5% of the speed of light, and maintain an acceleration of 1 g the whole time. This implies a mass ratio of 20 from table 11.1. But that's only for the journey out. To decelerate to a stop, we need another ratio of 20; to accelerate back toward Earth, another ratio of 20; and to decelerate to a stop again, another ratio of 20. The rocket equation is merciless: the overall mass ratio must be

$$R = 20^4 = 160,000.$$

In words, for every kilogram of mass on the spacecraft we need 160,000 kg of fuel, which is to say 80,000 kg of antimatter and 80,000 kg of normal matter. For our canonical 10,000 kg payload spacecraft (which, if manned, will probably need to be bigger), we need a whopping 1.6×10^{19} kg of fuel with a stored energy of 1.44×10^{26} J, or about as much energy as our current world civilization will use in about half a million years.

Be that as it may: the power the spacecraft will use on "takeoff" can be found from the fact that the power of a "photon rocket" such as this one is intimately related to the force the rocket exerts: if the force is F and the power P, then

$$F = 2P/c.$$

Since the rocket is accelerating at 1 g, and $g \sim 10 \,\text{m/s}^2$, the net force needed is 1.6×10^{10} N and the power required is 2.4×10^{18} W. This is an energy usage rate more than 100,000 times greater than our current civilization's. The total power the Earth receives from the Sun is about an order of magnitude less than this; if we launch from anywhere near Earth's orbit, we will likely destroy Earth's ecology, especially if we consider that (unlike the Sun) all of the energy is being delivered in the form of high-energy gamma rays [46, p. 138]. A starship like this one clearly cannot be launched from anywhere near Earth. Purcell also pointed out that at relativistic speeds like this, the hydrogen atoms that the ship intercepts essentially have the same energy as high-energy cosmic rays, meaning that extensive shielding would be needed to protect the crew from radiation exposure.

Purcell ended with this comment:

> Well, this is preposterous, you are saying. That is exactly my point. It *is* preposterous. And remember, our conclusions are forced upon us by the elementary laws of mechanics. All those people who have been talking about *lebensraum* in space and so on, simply haven't made this calculation and until they do, what they say is nonsense. [46, p. 138]

I am not sure it is nonsense, but it is clearly beyond anything our human civilization can do, and perhaps beyond what any civilization can do. Or perhaps not: in chapter 21, "A Googol Years," I'll discuss the issue of really advanced civilizations. In the meanwhile, however, we've talked about relativity a lot but haven't gotten to the good part: the issues of time dilation. Even though the universe is huge, given access to enough energy one could explore it within one human lifetime because of the theory of relativity.

NOTES

1. "Corpsicles" are people placed in cryogenic suspension just before death and revived, in different bodies, by the state hundreds of years later. This sort of cryonics is popular in science fiction. Perhaps the most widely seen use of it is in the TV show *Futurama*.

2. The author of this paper has suggested that because of *Bremsstrahlung* losses, using a ramscoop might be a good way of decelerating a ship traveling at a large fraction of the speed of light.

3. One must take care when reading this report to distinguish between two types of efficiency: overall energy efficiency, which is relatively low, versus the efficiency of a high-energy proton striking a target in producing antiprotons, which is relatively high.

CHAPTER TWELVE

INTERSTELLAR TRAVEL AND RELATIVITY

12.1 TIME ENOUGH FOR ANYTHING

> He looked me up and down and said wonderingly, "I knew intellectually
> you would not have changed with the years. But to see it, to realize it, is
> another thing, eh? 'The Picture of Dorian Gray.'"
>
> —ROBERT A. HEINLEIN, *TIME FOR THE STARS*

Robert A. Heinlein's novel *Time for the Stars* is essentially one long
in-joke for physicists. The central characters of the novel are Tom and
Pat Bartlett, two identical twins who can communicate with each other
telepathically. In the novel, telepathy has a speed much faster than light.
Linked telepaths, usually pairs of identical twins, are used to maintain
communications between the starship *Lewis and Clark* and Earth. Tom
goes on the spacecraft while Pat stays home; the ship visits a number of
distant star systems, exploring and finding new Earth-like worlds. On
Tom's return, nearly seventy years have elapsed on Earth, but Tom has
only aged by five [113].

I call this a physicist's in-joke because Heinlein is illustrating what is
referred to as the twin paradox of relativity: take two identical twins, fly

one around the universe at nearly the speed of light, and leave the other at home. On the traveler's return, he or she will be younger than the stay-at-home, even though the two started out the same age. This is because according to Einstein's special theory of relativity, time runs at different rates in different reference frames.

This is another common theme in science fiction: the fact that time slows down when one "approaches the speed of light." It's a subtle issue, however, and is very easy to get wrong. In fact, Heinlein made some mistakes in his book when dealing with the subject, but more on that later. First, I want to list a few of the many books written using this theme:

- *The Forever War*, by Joe W. Haldeman. This story of a long-drawn-out conflict between humanity and an alien race has starships that move at speeds near light speed to travel between "collapsars" (black holes), which are used for faster-than-light travel. Alas, this doesn't work. The hero's girlfriend keeps herself young for him by shuttling back and forth at near light speeds between Earth and a distant colony world [107].
- Poul Anderson's novel, *Tau Zero*. In this work, mentioned in the last chapter, the crew of a doomed Bussard ramship is able to explore essentially the entire universe by traveling at speeds ever closer to the speed of light [22].
- *The Fifth Head of Cerberus*, by Gene Wolfe. In this novel an anthropologist travels from Earth to the double planets of St. Croix and St. Anne. It isn't a big part of the novel, but the anthropologist John Marsch mentions that eighty years have passed on Earth since he left it, a large part of his choice to stay rather than return home [253].
- Larris Niven's novel *A World out of Time*. The rammer Jerome Corbell travels to the galactic core and back, aging some 90 years, while three million years pass on Earth [182]. [1]

There are many, many others, and for good reason: relativity is good for the science fiction writer because it brings the stars closer to home, at least for the astronaut venturing out to them. It's not so simple for her stay-at-home relatives. The point is that the distance between Earth and other planets in the Solar System ranges from tens of millions of kilometers to billions of kilometers. These are large distances, to be sure, but ones that can be traversed in times ranging from a few years to a decade or so by chemical propulsion. We can imagine sending people to the planets in times commensurate with human life. If we imagine more advanced propulsion systems, the times become that much shorter.

Unfortunately, it seems there is no other intelligent life in the Solar System apart from humans, and no other habitable place apart from Earth. If we want to invoke the themes of contact or conflict with aliens or finding and settling Earth-like planets, the narratives must involve travel to other stars because there's nothing like that close to us. But the stars are a lot farther away than the planets in the Solar System: the nearest star system to our Solar System, the triple star system Alpha Centauri, is 4.3 light-years away: that is, it is so far that it takes light 4.3 years to get from there to here, a distance of 40 trillion km. Other stars are much farther away. Our own galaxy, the group of 200 billion stars of which our Sun is a part, is a great spiral 100,000 light-years across. Other galaxies are distances of millions of light-years away.

From our best knowledge of physics today, nothing can go faster than the speed of light. That means that it takes at least 4.3 years for a traveler (I'll call him Tom) to go from Earth to Alpha Centauri and another 4.3 years to return. But if Tom travels at a speed close to that of light, he doesn't experience 4.3 years spent on ship; it can take only a small fraction of the time. In principle, Tom can explore the universe in his lifetime as long as he is willing to come back to a world that has aged millions or billions of years in the meantime.

12.2   WAS EINSTEIN RIGHT?

This weird prediction—that clocks run more slowly when traveling close to light speed—has made many people question Einstein's results.[2] The weirdness isn't limited to time dilation; there is also relativistic length contraction. A spacecraft traveling close to the speed of light shrinks in the direction of motion. The formulas are actually quite simple. Let's say that Tom is in a spacecraft traveling along at some speed v, while Pat is standing still, watching him fly by. We'll put Pat in a space suit floating in empty space so we don't have to worry about the complication of gravity. Let's say the following: Pat has a stopwatch in his hand, as does Tom. As Tom speeds by him, both start their stopwatches at the same time and Pat measures a certain amount of time on his watch (say, 10 seconds) while simultaneously watching Tom's watch through the window of his spacecraft. If Pat measures time Δt_0 go by on his watch, he will see Tom's watch tick through less time. Letting Δt be the amount of time on Tom's watch, the two times are related by the formula

$$\Delta t = \Delta t_0 / \gamma, \tag{12.1}$$

where the all-important "gamma factor" is

$$\gamma = \frac{1}{\sqrt{1 - \left(\frac{v}{c}\right)^2}}. \tag{12.2}$$

The gamma factor is always greater than 1, meaning Pat will see less time go by on Tom's watch than on his. Table 12.1 shows how gamma varies with velocity.

Note that this is only really appreciable for times greater than about 10% of the speed of light. The length of Tom's ship as measured by Pat (and the length of any object in it, including Tom) shrinks in the direction of motion by the same factor.

Even though the gamma factor isn't large for low speeds, it is still measurable. To quote Edward Purcell, "Personally, I believe in special relativity. If it were not reliable, some expensive machines around here would be in very deep trouble" [46, p. 134]. The time dilation effect has been measured directly, and is measured directly almost every second of every day in particle accelerators around the world. Unstable particles have characteristic lifetimes, after which they decay into other particles. For example, the muon is a particle with mass 206 times the mass of the electron. It is unstable and decays via the reaction

$$\mu^+ \rightarrow e^+ + \bar{\nu}_\mu + \nu_e \ (2.22 \times 10^{-6}\,\text{s}). \tag{12.3}$$

It decays with a characteristic time of 2.22 μs; this is the decay time one finds for muons generated in lab experiments. However, muons generated by cosmic ray showers in Earth's atmosphere travel at speeds over 99% of the speed of light, and measurements on these muons show that their decay lifetime is more than seven times longer than what is

Table 12.1
Gamma Factor as a Function of Rocket Velocity

v/c	γ
0	1
0.1	1.01
0.2	1.02
0.4	1.09
0.5	1.15
0.6	1.25
0.866	2
0.9	2.29
0.95	3.2
0.99	7.09
0.995	10.01
0.9995	31.63
$1-\delta, \delta \ll 1$	$\sqrt{2/\delta}$
1	∞

measured in the lab, exactly as predicted by relativity theory [233]. This is an experiment I did as a graduate student and our undergraduates at St. Mary's College do as part of their third-year advanced lab course. Experiments with particles in particle accelerators show the same results: particle lifetimes are extended by the gamma factor, and no matter how much energy we put into the particles, they never travel faster than the speed of light. This is remarkable because in the highest-energy accelerators, particles end up traveling at speeds within 1 cm/s of light speed. Everything works out exactly as the theory of relativity says, to a precision of much better than 1%.

How about experiments done with real clocks? Yes, they have been done as well. The problems of doing such experiments are substantial: at speeds of a few hundred meters per second, a typical speed for an airplane, the gamma factor deviates from 1 by only about 10^{-13}. To measure the effect, you would have to run the experiment for a long time, because the accuracy of atomic clocks is only about one part in 10^{11} or 10^{12}; the experiments would have to run a long time because the difference between the readings on the clocks increases with time. In the

1970s tests were performed with atomic clocks carried on two airplanes that flew around the world, which were compared to clocks remaining stationary on the ground. Einstein passed with flying colors. The one subtlety here is that you have to take the rotation of the Earth into account as part of the speed of the airplane. For this reason, two planes were used: one going around the world from East to West, the other from West to East [252]. This may seem rather abstract, but today it is extremely important for our technology. Relativity lies at the cornerstone of a multi-billion-dollar industry, the global positioning system (GPS).

GPS determines the positions of objects on the Earth by triangulation: satellites in orbit around the Earth send radio signals with time stamps on them. By comparing the time stamps to the time on the ground, it is possible to determine the distance to the satellite, which is the speed of light multiplied by the time difference between the two. Using signals from at least four satellites and their known positions, one can triangulate a position on the ground. However, the clocks on the satellites run at different rates as clocks on the ground, in keeping with the theory of relativity. There are actually two different effects: one is relativistic time dilation owing to motion and the other is an effect we haven't considered yet, gravitational time dilation. Gravitational time dilation means that time slows down the further you are in a gravitational potential well. On the satellites, the gravitational time dilation speeds up clock rates as compared to those on the ground, and the motion effect slows them down. The gravitational effect is twice as big as the motion effect, but both must be included to calculate the total amount by which the clock rate changes. The effect is small, only about three parts in a billion, but if relativity weren't accounted for, the GPS system would stop functioning in less than an hour [146, p. 68]. To quote from Alfred Heick's textbook *GPS Satellite Surveying*,

> Relativistic effects are important in GPS surveying but fortunately can be accurately calculated.... [The difference in clock rates] corresponds to an increase in time of 38.3 μsec per day; the clocks in orbit appear to run faster.... [This effect] is corrected by adjusting the frequency of the satellite clocks in the factory before launch *to 10.22999999543 MHz* [from their fundamental frequency of 10.23 MHz].

This statement says two things: first, in the dry language of an engineering handbook, it is made quite clear that these relativistic effects are so commonplace that engineers routinely take them into account in a system that hundreds of millions of people use every day and that contributes billions of dollars to the world's commerce. Second, it tells you the phenomenal accuracy of radio and microwave engineering. So the next time someone tells you that Einstein was crazy, you can quote chapter and verse back at him!

12.3   SOME SUBTLETIES

The problem with Heinlein's *Time for the Stars* is that when the *Lewis and Clark* begins to get close to the speed of light, the twins have problems communicating with each other. This is Tom speaking:

> At three-quarters of the speed of light [Pat] began to complain that I was drawling while it seemed to me he was beginning to jabber. At nine-tenths of the speed of light it was close to 2 to 1, but we knew what was wrong now and I talked fast and he talked slow.
>
> At 99% of *c* it was 7 to 1 and all we could do to make ourselves understood. Later that day we fell out of touch entirely. [113, chap. 11, "Slippage"]

This seems reasonable, but unfortunately, it violates one of the underlying principles of the special theory of relativity. Relativity is called relativity because measurements made by an observer are relative to that observer and not to anyone else. The odd thing is that Tom will measure his own time going by normally and *Pat's clock slowed down—* by the same gamma factor that Tom's clock appears slowed down as measured by Pat. This is so odd that when I was teaching this once, one of my students shook her head and stated flatly, "No, that isn't right." This is one of the two underlying ideas of Einstein's special theory of relativity: you can't tell whether you are moving at constant velocity or standing still by any measurement you can make. *If* it were true that Pat would measure Tom's clock as running slow if Tom were moving and he weren't, and Tom would measure Pat's clock as running fast, then that would be proof that Tom was moving and Pat wasn't. Since you can't do that, both clocks must run slow as measured by the other.[3] (How this works and makes sense would take far too long to go into in this book. If you are interested, there are good books on the subject, a few written by Einstein himself [198].) The other principle of relativity, which is what leads to the time dilation effects, is that both Tom and Pat will measure the speed of light as having the same value no matter how fast they are moving in relation to each other.

Ultimately the problem with Heinlein's novel comes from the fact that the twins are communicating with each other at speeds faster than light, which violates the precepts of the special theory of relativity. If we put $v > c$ into the equation for the gamma factor we get the square root of a negative number, that is, an imaginary quantity—indicating that we can't do it. Some people have tried to come up with clever ways around this problem, but most physicists think that the speed of light is the ultimate limiting speed in the universe.

This does lead to another issue: why is it that at the end of the voyage, Tom is younger than Pat? If the effect is symmetric, why should either of them be younger than the other?

This has been discussed to death in the past. The answer is straightforward: yes, time dilation is symmetric, but there is something that isn't. That is their *accelerations*. Pat, sitting still on Earth, is not being accelerated. Tom, on the other hand, when he climbs aboard the ship and takes off for the stars, is. *Even if the acceleration is for a very short period of time, this difference is what makes Tom younger than Pat.* Let me work through an example to show why this is true. Here are the assumptions I'm going to make. This example is based on the discussion in Wolfgang Rindler's book, *Special Relativity* [198, pp. 30–31].

- Tom is going to make a trip to the Alpha Centauri star system, 4.3 light-years away.
- His spacecraft is capable of traveling at 86% of the speed of light after a very rapid period of acceleration. (We'll ignore the problem of keeping Tom from getting smashed to jelly during the acceleration.)

At 86% of the speed of light it takes Tom five years to get there and five years to get back *as measured by Pat's clocks.* The gamma factor is almost exactly 2, however, so to Tom it should take only 2.5 years out and back, meaning he should be five years younger than Pat on his return. This is from Pat's perspective. How about from Tom's perspective?

A rather incisive point made by Rindler is that no matter how fast the acceleration period is, when it is over Tom has gone halfway to his destination [198, p. 31]. This is a result of the length contraction effect: from Tom's point of views when the acceleration ends, the Sun is moving away from him at 86% of the speed of light and Alpha Centauri is moving toward him at that speed. Because of relativistic length contraction, the distance between them has shrunk by 50%. To Tom, his clock is normal, but the distance is contracted by 50%. So it all works out.[4]

12.4 CONSTANT ACCELERATION IN RELATIVITY

There is no such thing as constant acceleration in special relativity, for a simple reason: under constant linear acceleration (with initial condition $v = 0$ at time $t = 0$), the velocity after time t is $v = at$. Because of this, after a sufficiently long time (about one year if $a = g$), the spacecraft is traveling faster than the speed of light. We therefore must be careful about how we define acceleration.

An astronaut on an accelerating spacecraft will feel the sensation of weight, so this is what we will adopt as our definition of acceleration.

1. Weigh the astronaut on Earth (W_0).
2. Then, once on the spacecraft, put a scale under her and measure her effective weight (W).

The "proper" acceleration of the spacecraft, a, is then

$$a = g \frac{W}{W_0}.$$

Since $g \sim 10 \text{ m/s}^2$, if the weight aboard the craft is only 10% of the weight on Earth, the acceleration of the spacecraft is about 1 m/s^2.

It turns out that we can make life very simple for ourselves in talking about acceleration if we adopt a system of units in which distances are measured in light-years and time is measured in years. That is, the speed of light is $c = 1$ LY/yr. In this system of units, g works out almost exactly to 1 LY/yr^2, which makes our calculations very easy. Let's assume the following:

- The ship starts from rest and travels out with a constant acceleration of g along a straight line.
- x is the distance the ship travels.
- t is the time from the beginning of the trip as measured on clocks on Earth.
- v is the speed the ship gets to after time t.
- τ is the time that has passed on the ship.

Then the motion of the ship follows the equation

$$(x + 1)^2 - t^2 = 1. \tag{12.4}$$

This is often referred to as "hyperbolic motion" in relativity because if x is plotted against t, the figure is that of a hyperbola.

We can express everything in terms of the time on board the ship:

$$x = \cosh(\tau) - 1, \tag{12.5}$$

$$t = \sinh(\tau), \tag{12.6}$$

$$v = c \tanh(\tau). \tag{12.7}$$

For those unfamiliar with the terms, "cosh," "sinh," and "tanh" are the hyperbolic cosine, sine, and tangents, respectively:

$$\cosh(x) = \frac{e^x + e^{-x}}{2},$$

$$\sinh(x) = \frac{e^x - e^{-x}}{2},$$

Table 12.2
Time and Space under Relativistic Acceleration

τ_{end} (yr)	x_{end} (LY)	t_{end} (yr)	v_{mid}/c
0.5	0.06	0.5	0.24
1	0.26	1.04	0.46
2	1.08	2.36	0.76
3	2.7	4.26	0.91
4	5.52	7.26	0.96
5	10.26	12.1	0.99
10	146	148	$1-\epsilon$
20	22000	22000	...
40	4.85×10^8	4.85×10^8	...
50	7.2×10^{10}	7.2×10^{10}	...

and

$$\tanh(x) = \frac{\sinh(x)}{\cosh(x)} = \frac{e^x - e^{-x}}{e^x + e^{-x}}.$$

12.4.1 A Trip to the End of the Universe

Let's plan a trip to the great beyond. After all, a trip of a billion light-years begins with a single step, right? The nice thing about relativistic acceleration is that although nothing can go faster than the speed of light. It will take at least a billion years to go a distance of a billion light-years, but time dilation makes it seem much, much shorter for the voyagers. Here's the idea: we'll take our ship and accelerate for half the time (i.e., half the distance) of the trip (as measured by shipboard clocks), turn the ship around, and decelerate the other half. How far do we get in this time? How much time has elapsed on Earth? How fast were we going at midpoint? In table 12.2, x_{end} is the distance the ship travels by the end of the voyage, t_{end} and τ_{end} are the times as measured on Earth and on the ship's clock, respectively, and v_{mid} is the speed at midpoint.

The results are clear: given a ship capable of accelerating at $1\,g$ continuously, one could reach the nearest stars in a few years, the center of the galaxy in under 25 years, other galaxies in 40 years, and the edge of the universe in 50 years. So the universe is within our grasp! Unfortunately, the energy requirements for a trip like this mount up considerably, but I'll leave this as an exercise for my readers to work through.

NOTES

1. One point concerning this novel is that much of the time dilation experienced by Corbell is the result of diving near the event horizon of the supermassive black hole at the center of the galaxy. Gravitational time dilation is from the general theory of relativity, which isn't covered in this chapter.

2. Indeed, the website Crank Dot Net, which specializes in providing links (I quote) to "cranks, crackpots, and loons on the net," lists no less than 48 sites dedicated to showing that the theory of relativity (really, the two theories of relativity, special and general) are wrong [3].

3. I am being careful to avoid saying "it *appears* that clocks run slow" because this implies that the rate changes are somehow an illusion. This is wrong. By any test you can make, a clock moving in relation to you runs slow.

4. The readings on Pat's clocks are harder to explain, but one can show that there are more ticks on Pat's clock than on Tom's using a combination of time dilation and what is called the *Doppler effect* from Tom's point of view.

CHAPTER THIRTEEN

FASTER-THAN-LIGHT TRAVEL AND TIME TRAVEL

> Regular space is deep and wide,
> Hyperspace is just outside.
> —THE SPACE CHILD'S MOTHER GOOSE

13.1 THE REALISTIC ANSWER

Faster-than-light travel and time travel are both impossible.

13.2 THE UNREALISTIC ANSWER

I'm in a somewhat difficult position writing this chapter, for several reasons:

- I don't believe that the laws of physics allow faster-than-light travel or time travel into the past.

- However, both of these ideas are strong elements in many science fiction stories.
- Compounding the problem, there is a plethora of recent books on the subject, and almost anything I can say about the subject has already been said by those authors, and better.

Faster-than-light (FTL) travel is a key component of many science fiction stories, for obvious reasons: traveling to the stars is an undertaking that requires enormous resources and takes a very long time even if we push light speed in doing so. The distances between the stars are just not commensurate with human length scales or time scales. There are a number of different approaches to getting around this problem:

- Ignore relativity entirely. If we assume that relativity is wrong, then we can go as fast as we like. No one likes this approach much today, as there is overwhelming evidence that relativity works. The original *Star Trek* series seemed to ignore relativity, although I realize that the mere act of writing this down will have some fan quoting chapter and verse to contradict me. The surrealistic TV series *Space:1999* certainly ignored it, although this wasn't its biggest break with reality or common sense by a long shot. E. E. "Doc" Smith in the Lensman series had an "inertialess" drive that somehow worked around relativity to allow ships to be accelerated to near-infinite speeds [221]. This isn't so much ignoring relativity as bypassing it, but it's still impossible.
- Use tachyons. Tachyons are completely hypothetical particles that have never, ever been seen and that in principle travel faster than the speed of light [80]. They almost certainly don't exist. Robert Erlich is one physicist who thinks they do. He thinks that neutrinos might be tachyons. He's probably wrong, but his ideas are interesting [79].
- Use some idea concerning "quantum entanglement." This is probably what Ursula K. Le Guin based her idea of the "ansible communicator" on [145]. It stems from the fact that in quantum mechanics, the "collapse" of a wave function for two "entangled" particles happens instantaneously—at least in some reference frame. This doesn't violate the theory of relativity, however, because (at least according to the Copenhagen interpretation of quantum mechanics) wave function collapse cannot transmit information or energy.[1] So much for the ansible.

- Use hyperspace. Hyperspace is some "multidimensional" analog to real space; how it works is a little vague. I think the idea is that the speed of light is effectively infinite in hyperspace and that there is a point-to-point correspondence between hyperspace and regular space, so that by putting a ship there and moving around a bit, you emerge back in regular space light-years away. Books using this idea in various forms include *The Mote in God's Eye* and its sequels, Larry Niven's *Known Space* stories, the TV show *Babylon 5*, and possibly the reimagined *Battlestar Galactica* as well.
- Use folded space-time and wormholes. In the general theory of relativity, space behaves in some ways (in higher dimensions) like a rubber sheet distorted by the masses placed in it. There is nothing to prevent the rubber sheet from being bent so that two points, although far away on the sheet, are close in a direction perpendicular to the sheet (as seen from a higher-dimensional perspective). Some sort of bridge between those points would allow a ship to travel from one to the other in a much shorter time than it would take to move along the sheet. The first place I read about this was in Madeleine L'Engle's *A Wrinkle in Time*, in which the author incorrectly described this means of transportation as a "tesseract" [152][2]. This is just about the only idea that has any scientific merit; the relativist Kip Thorne investigated whether this would work for FTL travel at the request of his friend Carl Sagan for use in the novel *Contact* [209]. Wormholes are the practical way to exploit folded space, as they form the aforementioned bridges.

Kip Thorne has discussed using wormholes for effectively FTL travel extensively, so I will just summarize here. I do need to discuss some ideas from Einstein's theory of relativity to introduce the subject.

13.3   WHY FTL MEANS TIME TRAVEL

It is a fairly common statement that FTL travel implies time travel into the past. However, the reasons for this aren't often discussed, as it takes a fairly subtle understanding of the theory of relativity to appreciate why this should be so. What is also true, at least as far as the special theory of relativity is concerned, is that allowing matter or energy to travel faster than light results in some bad paradoxes—real ones, not seeming paradoxes like the twin paradox discussed in the last chapter but ones that cannot be resolved without abandoning one of the two principles of relativity. It isn't clear whether the general theory of relativity resolves these paradoxes and allows FTL travel and time travel. For now, let's discuss things in the context of special relativity.

First some definitions from the theory of relativity. An *event* is anything that happens, plus the time and place where and when it happens. Generally speaking, we reserve the term for things that are of short duration so that we can define the time with reasonable precision. Let's take two events; with complete lack of imagination I'll call the first one A and the second one B. Let's say A is lighting a fuse and B is a firecracker going off. Did A cause B?

This is an interesting question. Let's put a long fuse on the firecracker, say five feet long, which burns at a rate of 1 ft/s. If the firecracker explodes six seconds after the fuse is lit, then lighting the fuse may have caused the firecracker to explode. If it explodes four seconds after, then A couldn't have caused B because there wouldn't have been enough time for the fuse to burn down. However, we can (in principle) make a fuse that burns faster.

Let's make a really long fuse: five light-seconds long, or roughly 1.5 million km. We'll light the firecracker using light—imagine that we have some sort of optical fiber, a laser to send the signal, and a detector on the firecracker. The signal travels at a speed of one light-second every second, so it takes five seconds for the light to reach the firecracker. Again, if the cracker explodes in four seconds, then sending the signal could not have caused the explosion; if it explodes after six seconds, then it could have. The point here is that according to all we know about physics, we cannot increase the signal speed beyond the speed of light. If the firecracker explodes four seconds after the light signal was sent, there is no possible way that A could have caused B.

What is interesting is that in that case it doesn't really matter that A occurred before B. Philosophically speaking, if A cannot possibly influence B, it doesn't matter when they happen relative to each other. Amazingly, this philosophy is embedded in the mathematics of the theory of relativity! As we saw in an earlier chapter, time flows at different rates for people moving relative to one another. If A cannot possibly cause B, then in some reference frames A will precede B, in others B will precede A, and in some both A and B will happen at the same time. This is known as the "relativity of simultaneity," and is one of the most subtle and interesting things about the special theory.

Now lets modify things so that the fuse burns with a speed faster than light—say, twice the speed. Then the firecracker will explode 2.5 seconds after the fuse is lit. The problem is, in which reference frame? In one reference frame the firecracker explodes after the fuse is lit, in another before the fuse is lit, and in a third at exactly the same time. What this means is that if we accept that the special theory of relativity works, we cannot assign a consistent time ordering to these events. For those interested in reading about this in more detail, I recommend the two books *It's about Time* by N. David Mermin and *Spacetime Physics* by Edwin F. Taylor and John Archibald Wheeler, in particular the section on Lorentz transformations and the parable of the "great betrayal" [165][234].

The special theory *does* work in regions of space that are free of large masses, such as the vast gulfs between the stars. There is overwhelming proof that we can use the postulates and the results of the special theory; trillions of particles generated in accelerators all over the world can't be wrong, can they? The conclusion has to be that a spaceship simply can't travel faster than the speed of light through normal space.

Science fiction writers who include things that travel faster than light in their stories generally do so by the implicit assumption of a preferred reference frame from which times and distances are measured. This is usually time and space as measured by clocks on Earth. It is the assumption that Robert Heinlein makes *Time for the Stars*, when Tom begins to sound slow to Pat and Pat fast to him as his ship approaches the speed of light [113]. It may also be true in a more subtle way for the Alderson drive used in *The Mote in God's Eye*. The following quotation is from the essay "Building the Mote in God's Eye," in which Niven and Pournelle describe how one uses the drive in the novel:

> There are severe conditions to entering and leaving the continuum universe [i.e., hyperspace]. To emerge from the continuum universe you must exit with precisely the same potential energy . . . as you entered. You must also have zero kinetic energy relative to a complex set of coordinates that we won't discuss here [186].

I remember reading this essay in *Galaxy* magazine when I was about ten years old and wanting to know more details of how it worked. It seems that the Alderson drive implicitly violates the special theory by imposing a privileged reference frame from which times, distances, and energies are measured. Again, the ships disappear in one place and reappear in another, light-years away, "instantly," but again, there are reference frames in which the ship will reappear before it disappears.

Physicists like causality: the idea that if event A causes event B, A must unambiguously happen before B does. Therefore, directly traveling faster than the speed of light is out. Is there any other way?

13.4   THE GENERAL THEORY

If faster-than-light travel is allowed in our universe, we have to appeal to the general theory of relativity for this. The general theory of relativity expresses how space and time are "warped" (curved) in the presence of mass, and how this spatial curvature affects the trajectories of particles moving near the masses.

One of the common metaphors used in discussing this subject is to say that space is something like a giant stretched sheet of rubber and the stars are bowling balls dropped onto the sheet. If we were to roll a marble across the sheet near the bowling ball it would be deflected from a straight line because of the depression created by the bowling ball, not because of some "attraction" between the bowling ball and the marble. The marble represents a planet or a spaceship on a trajectory flying close to the star. It's not a great analogy because we are dealing with the curvature of time and space, not just space; clocks on the spacecraft will be slowed as they approach the star, and so will show a time difference compared to ones that are farther away.

This deformation of space-time is what would, hypothetically, make it possible to take a "shortcut" around normal space. I'm going to be more descriptive in this chapter than in others because the mathematics one needs for understanding the general theory is very difficult. Where possible I will put the numbers in.

One of the first tests of the general theory of relativity was the deflection of light by a large mass, in this case the Sun. In 1919 Sir Arthur Stanley Eddington photographed stars that were near the Sun during a total eclipse and found that their positions were changed from their apparent positions six months later, when they were far away from it. This is not surprising: one might expect Newton's older theory of gravity to predict this as well. Light rays from a distant star skimming by the Sun would be attracted by the strong gravitational "pull" of the star, and would therefore be bent, making the star appear in a different place in the sky. However, Newton's theory predicts a different value than Einstein's theory: deflection under the Einsteinian theory is twice what Newton predicts. If anyone is interested in using this in a story, the angular deflection of a light ray skimming by the edge of a star like our Sun (in radians) is

$$\theta = 4\frac{G\,M_{metric}}{c^2\,R_{metric}}. \tag{13.1}$$

Here, $G = 6.67\times10^{-11}$ Nm^2/kg^2 is the universal gravitational constant, and M_{metric} and R_{metric} are the mass and radius of the star. The subscript "metric" means that in this equation we use the metric value for these units: for the Sun, the metric mass is 1.99×10^{30} kg and the metric radius is 6.85×10^8 m. In this formula the deflection is expressed in radians, where 1 rad $\approx$ 57.3 degrees. In more useful units,

$$\theta = 1.77'' \times \frac{M}{R}, \tag{13.2}$$

where M and R are relative to the mass and radius of the Sun (i.e., $M_{sun} = 1$ and $R_{sun} = 1$ in these units), and 1 arc-second ($''$) is 1/3,600 of a degree. These are units we will use a lot in later chapters when discussing the subject of life in the universe.

The deflection of light even due to a massive object like the Sun is small. This is because the amount of space-time curvature around the Sun is small. The curvature of space-time around a mass of a given radius is a function of the dimensionless parameter:

$$\frac{G\,M_{metric}}{c^2 R_{metric}} = 2.15 \times 10^{-6} \times \frac{M}{R}. \tag{13.3}$$

If this number is small, the effects due to the general theory of relativity (curvature of space and time dilation) are small, but as it approaches 1 the curvature becomes severe. If it is greater than 1/2, the star disappears from the universe entirely.

13.5   GRAVITATIONAL TIME DILATION AND BLACK HOLES

The escape velocity of a rocket taking off from a planet of radius R is given by the formula

$$v_{esc} = \sqrt{\frac{2G\,M_{metric}}{R_{metric}}}. \tag{13.4}$$

If the rocket travels at less than this speed, it eventually falls back to the planet. If it has an initial speed equal to or greater than this speed, it doesn't come back down. Nothing travels faster than the speed of light, so if we insert light speed into the equation and solve for the radius, we find the following:

$$R_{metric} = \frac{2G\,M_{metric}}{c^2} = 2.9\,\text{km} \times M, \tag{13.5}$$

where the second expression is in units where the Sun's mass is 1. That is, if we squeezed all of the Sun's mass into a radius of less than 2.9 km, it would become dark: light itself couldn't escape from its gravitational pull. This is using the Newtonian formula, but in fact, Einstein's theory of relativity predicts the same thing: one must interpret the radius in a somewhat different way because of space-time curvature, but the formula is the same. For a good, elementary discussion of general relativity I highly recommend the book *Exploring Black Holes*, by Edwin F. Taylor and John Archibald Wheeler [135]. This is almost a handbook for science fiction writers trying to use black holes in their stories in some way. The level of the book is about that of a senior in college majoring in physics.

Black holes exist: they are the product of the end stages of the evolution of massive stars, ones more than twenty times the mass of the Sun. Stars are a balance between the heat and pressure pushing outward generated by fusion at their cores and the force of gravity pulling them in. Once the fusion fires are ended, if the star is large enough nothing can prevent it from collapsing completely and effectively disappearing from our universe. What is left behind is an "event horizon," a one-way barrier with radius given by equation (13.5). Once you cross it you cannot return to the rest of the universe. Beyond that point one must inexorably fall to the singularity at the center, where tidal forces inevitably destroy everything entering the black hole's maw. What happens to matter once it falls to the singularity isn't very well understood; understanding it requires a complete quantum theory of gravity, which we don't have yet.

Another fascinating aspect of black holes is that from the view of an outside observer, time slows down as you approach the event horizon. At the horizon, it slows to a stop. However, someone falling through the horizon to the singularity notices nothing unusual! Imagine we have two people, one (Al) in a spaceship at a very large distance away from the black hole and another (Bert) in a spacecraft closer to it. Assume both are at rest with respect to the black hole (i.e., their spaceships are firing their engines to keep from falling into it). The clock on Al's spaceship will run faster than the one on Bert's. If Al measures a time t_A on his clock, he will see time t_B run by on Bert's clock, where

$$t_B = \sqrt{1 - \frac{2G\,M_{metric}}{c^2 r_{metric}}} \times t_A \qquad (13.6)$$

on Bert's clock in that time. Here, r is the distance of Bert from the center of the black hole [235]. If either observer is moving relative to the center of the black hole, the formula must be changed to account for the motion. As expected, time on Bert's clock slows to zero as one approaches the event horizon. As I mentioned in chapter 12, this gravitational time dilation and the relativistic motion effects must be taken into account for the GPS system to work. This to me, is one of the most amazing effects of relativity. Time slowing as one approaches the event horizon has been used in a number of science fiction stories, not always accurately.

Black holes have fascinated science fiction writers nearly as long as they have fascinated physicists. As Kip Thorne says, the "golden age" of black hole research was in the 1960s, and it is in literature from the 1960s and early 1970s that we find black holes first making their way into mainstream science fiction. It's possible that the "black sun" of Arthur C. Clarke's *The City and the Stars* has an early use of the idea. In that story the black sun, an artificially constructed star, is used to imprison the "Mad Mind," a noncorporeal being created by the denizens of a galactic empire, which turned on them and nearly destroyed the galaxy [53]. If so, Clarke was being prescient, as the book predicted that eventually the Mad Mind would escape once the black sun failed. When the novel was written, in the late 1950s, it wasn't known that black holes evaporate through Hawking radiation. Eventually (long, long after all the stars are dead) all of the black holes will boil away to nothingness.

The story "He Fell into a Dark Hole" by Jerry Pournelle features the rescue of a group of spacemen stranded in orbit around a black hole, and the ship being nearly destroyed by gravitational radiation emitted by the hole. The story is set in the same universe as *The Mote in God's Eye* and is scientifically accurate except for the idea that astrophysicists would have forgotten all about black holes five centuries from now. Larry Niven, of course, wrote several stories featuring black holes, including "Singularities Make Me Nervous," which involves an astronaut who travels backward in time by going through a black hole.

Black holes, occasionally called "collapsars," have been used for FTL travel in a number of science fiction novels. Two that spring readily to mind are *The Forever War* by Joe W. Haldeman and *Fiasco* by Stanislaw Lem [107][151]. I've mentioned the first book already. The second records the attempt of a human expedition to make first contact with an alien race, the Quintans. In the book the ship uses a "gracer" (a gravitational laser, whatever that is) to make the collapsar oscillate, permitting the ship to dive through it and travel either faster than light or backward in time—the rather recondite language of the book doesn't really make it clear which. Lem was a highly educated man, and so this bit probably came from work physicists had done on so-called Kerr black holes.

The relationship between time travel, FTL travel, and black holes has to do with the Kerr solutions of the Einstein field equations for rotating black holes. At the center of every black hole is a singularity, a place where space is infinitely curved, where anything falling into it is torn apart by simultaneous squeezing and compression along different axes. For a nonrotating hole, the singularity is a point, meaning that anyone falling through the singularity will be killed, but a rapidly rotating black hole has a singularity in the form of a ring. The solutions of the Einstein equations for an object falling through the ring are very strange: they indicate that an astronaut falling through the singularity could travel along a "closed timelike curve," which is to say into his past. They also appear to permit FTL travel as well. Unfortunately, the solutions are unstable, and any matter entering the ring singularity almost certainly destroys it, killing the traveler in the process [168]. This unfortunately scotches Niven's and Lem's stories and movies like *The Black Hole* in which people travel beyond the singularity.

13.6   WORMHOLES AND EXOTIC MATTER

A number of solutions of Einstein's equations lead to odd, seemingly unphysical behavior. Kurt Gödel, the great mathematician, found equations describing a universe in constant rotation in which travel indefinitely into the past and future was possible by picking the correct trajectory to follow. Because the real universe is not in rotation, this isn't taken very seriously, but in 1974 Frank Tipler showed that closed timelike curves appeared around an infinitely long, rapidly rotating cylinder. The paper is titled "Rotating Cylinders and the Possibility of Global Causality Violation" [237]; Tipler showed that if the outer edge of the cylinder rotated at speeds greater than half the speed of light, it could be used as a time machine.

It is unclear whether a long but finite, rapidly spinning cylinder could do the same thing. Tipler thought so, but other physicists have pointed out some unphysical properties of this solution of Einstein's equations. Tipler's paper has been used in at least two science fiction stories written in the late 1970s. In the first, Larry Niven's "Rotating Cylinders and the

Possibility of Global Causality Violation," various alien races try to build a Tipler cylinder, only to be stopped by one natural disaster after another [184]. In the final paragraph, as the scientist is telling his monarch about the device, the star their planet circles goes nova. This is the universe protecting itself from anyone building a time machine. It is a radical example of Steven Hawking's "chronology protection hypothesis," the idea that the universe will not permit the existence of a time machine. It was originally a science fiction idea, but modern research in gravitational physics has explored this idea in some detail. The other story is the novel *The Avatar*, by Poul Anderson, which is a more conventional space travel story in which the Tipler device (called "T-machine" in the text) is used for faster-than-light travel [21].

In 1988 Kip Thorne, a Caltech physicist, was asked by his friend Carl Sagan to look over the manuscript for the novel *Contact* and critique it for scientific accuracy. In the first draft the heroine, Ellen Archway, traveled to a distant star by diving through a black hole. As I have said, this doesn't work: Ellen would have been killed by the singularity at the center. To help Sagan out, Thorne investigated whether another solution of Einstein's equations, the wormhole, was a possibility. Wormholes form a bridge (more or less a tube) between two different regions of space-time, allowing travel between two widely separated parts of the universe. This does require two different things happening, however:

1. Space must be bent like a crumpled sheet of paper as seen through higher dimensions, or (if it is not already folded) the sufficiently advanced civilization must be able to fold space in this manner. This idea is not uncommon in science fiction: Heinlein's novel *Starman Jones* posits interstellar travel using this trick. In it, the astrogators of the starships pilot their craft through "Horstian anomalies," places where distant points in space are folded together through higher dimensions[112]. To quote Max, the protagonist of the story, showing off for his girlfriend:

   > "You can't go faster than light, not in our space. If you do, you burst out of it. But if you do it where space is folded back and congruent, you pop back in our space—but a long way off. How far off depends on how it's folded."

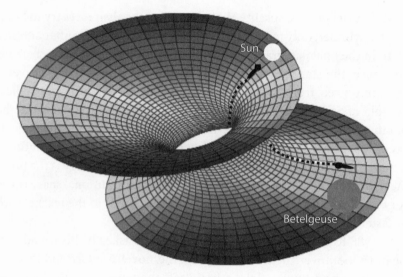

Figure 13.1. Space-time diagram of a wormhole.

Ships travel to near the speed of light and then reach it and exceed it at the anomaly point, "bursting through" to points light-years away. The plot centers on a ship getting lost when the astrogators make a bad calculation right before the "jump." Heinlein is again making the error of assuming a privileged reference frame: the ship's speed depends on who is looking at it, and no object with mass can move at, let alone exceed, the speed of light. The book, written in 1953, is still a fun read despite anachronisms such as human astrogators programming computers in binary code in 3000 CE.

2. If space is folded or can be folded in this way, one must also be able to create and maintain the wormhole. Morris and Thorne found out that this would require the wormhole to be threaded with *exotic matter*.

If one looks at standard diagrams for wormholes, they simply *look* like they need to be held apart by something, some sort of bands or hoops running through them to keep them from collapsing back on themselves (fig 13.1).

What Thorne and Morris found was that these bands needed to be of a strange form of mass or energy that had a tension greater than its energy density. One can kind of rationalize this by saying that the mass of the exotic matter would tend to collapse the wormhole owing to its gravitational self-attraction, while the tension would be needed to keep it from collapsing, so the tension must be greater, but this is a gross oversimplification. Morris and Thorne called this material "exotic" because it is very different from any matter we have experience with. In an earlier chapter we discussed one example of a structure built entirely in tension, the space elevator. A wormhole is another structure entirely in tension, but a much more extreme one. Let's define a dimensionless parameter, the "exoticity" of a given material, as

$$\eta = \frac{Y_{max}}{\rho c^2} = h \times \frac{g}{c^2}. \tag{13.7}$$

The smaller η is, the more "normal" the material is. The value h is the "breaking height" of a structure as defined in chapter 5, the maximum length of a cable made of the material that can be supported under gravity. Carbon nanotubes, the materials with the highest strength-to-weight ratio known, and therefore the ones with the highest exoticities, have $h \approx 10^6$ m, or an exoticity parameter of 10^{-10}. Puting it differently, since the acceleration of gravity is just about 1 LY/yr^2, the exoticity parameter can be written in a very simple form:

$$\eta = h \, (\text{LY}). \tag{13.8}$$

It is essentially equal to its breaking length *expressed in light-years*. To construct a wormhole, this parameter must be greater than 1. Another way to say it is that to create a wormhole, one would need something that could support a length of itself at least one light-year long against Earth's gravity. There is a generic proof that any means of traveling faster than light or traveling backward in time must involve exotic matter in its creation [168].

Exotic matter reminds me of the chain *gleipnir* used to bind the Fenris Wolf in Norse mythology: made from the sounds of cats' paws, the breath of fish, the spittle of birds, the hairs of a woman's beard, and the roots of a mountain, it was as thin as a thread but strong enough to hold the fiercest monster bound fast [62]. It seems to me that this would work as well as anything else in keeping a wormhole throat open. I therefore propose renaming exotic matter *gleipnirsmalmi* ("gleipnir's metal," about the closest I could come to the idea using an online Icelandic dictionary).

Physicists have proposed things almost as exotic as the ingredients used in the myth to keep wormholes open. It's relatively easy to show that ordinary solid matter cannot be exotic. One of the web problems for this book works this constraint through in detail. Exotic matter is called exotic because it violates the weak energy condition of general relativity; in some reference frame the mass density will be negative [168][169][236]. The reason is that a parallel bundle of light rays going through the wormhole will be defocused; gravitational lensing by ordinary matter focuses light, meaning that in some ways, this *gleipnirsmalmi* behaves almost like antigravity.[3] One candidate is the Casimir vacuum energy; in quantum mechanics the vacuum isn't really empty but is filled with particles and fields popping in and out of existence; as long as this happens in a time too short for anyone to notice, there's no problem. The Casimir vacuum is achieved by closing off a piece of space, like putting two highly reflecting parallel plates a distance *a* apart from each other. Hendrik G. B. Casimir showed in 1948 that this arrangement lowered the energy of the vacuum between them, resulting in an attractive force [48]

$$ F = \frac{\pi^2 h c A}{1440 a^4}. \tag{13.9} $$

The Casimir energy is exotic; the exoticity parameter $\eta = 3$. Unfortunately the classical Casimir effect between two plates probably doesn't work. In the original derivation, Casimir used a mathematical trick, essentially cutting off the calculation at high energies. The effects of vacuum modes above the cutoff point and of the mass of the plates themselves cannot be ignored, meaning that attempting to use the Casimir effect due to real, physical metal plates is not going to work [242, pp. 123–124].

How much exotic matter do we need to produce a wormhole? The mathematics needed to answer this question goes beyond the scope of this book. Matt Visser made the following estimates for the tension in the wormhole throat and the total "mass" of exotic matter needed.

These are

$$M = 10^{27} \text{ kg} \times b \text{ (meters)} \tag{13.10}$$

and

$$\tau = 5 \times 10^{42} \text{ N/m}^2/b^2 \text{ (meters)}, \tag{13.11}$$

where b is the throat radius, τ is the tension, and M is the "mass" of the exotic matter. One point that Visser discusses is that M isn't exactly mass because of the strange nature of exotic matter and space-time curvature. However, it is a measure of the energy content of the exotic matter we need. A one-meter-diameter wormhole will require a quantity of exotic matter equivalent to the mass of all of the planets in the Solar System. One meter probably won't do it, however: throats that small would probably tear apart anyone going through them because of tidal forces. The tension is far, far greater than any possible material can sustain.

One really interesting feature of wormholes used for interstellar travel is that their mass depends on the mass of whatever passes through them. I don't think that any science fiction writer has ever used this point in any story. The issue is this: the conservation laws of physics are local in nature. That is, we think that mass and energy are conserved. However, they are conserved locally: you can't simply have 10,000 kg disappear in one place (say, near the Sun) and 10,000 kg reappear in another place (say, near the star Betelgeuse, 600 light-years away) at the same time. Why not? Because "at the same time" is a relative statement: in one reference frame, they will disappear and reappear at the same time. In another, the mass will disappear and there will be an interval before it reappears. In another, the mass will reappear near Betelgeuse before disappearing. Richard Feynman stated the law thus: if you have a certain amount of mass-energy inside a box, the only way the amount inside the box can change is if you move a mass through the walls of the box [81, pp. 63–65].

Let's consider the two mouths of the wormhole. Put a box around each mouth. A 10,000 kg spacecraft goes through the mouth of the wormhole near the Sun and reappears out of the other mouth near Betelgeuse. Well, 10,000 kg just went through the box near the Sun and didn't come out again, according to an observer near the Sun. The wormhole mouth near the Sun just gained 10,000 kg, if we believe the conservation of mass-energy. An observer near Betelgeuse just saw 10,000 kg emerge from the wormhole mouth near that star. The mouth near that star must have lost 10,000 kg. This can be rigorously justified using the general theory of relativity [242, p. 111]. In his book, Visser raises the interesting and unanswered question: what happens if one mouth loses so much that its mass becomes negative?

The total charge of a system is conserved in the same way mass is. If a positive charge goes through the wormhole, the field lines from the charge still stick out of the mouth it entered. This makes it "look like" a positive charge to an observer at the mouth near the Sun. When the positive charge emerges from the other mouth, the field lines from the charge will be bunched up as they pass through the second mouth, making the mouth near Betelgeuse "look like" a negative charge. John Wheeler once proposed that the reason why the universe is charge neutral is that there really is no such thing as charge on the most basic level: charges are really electric field lines threading the twin mouths of wormholes. The idea probably doesn't hold up, but it is pretty neat.

Can one use wormholes for time travel? Yes. Kip Thorne showed that if you took one mouth of the wormhole and accelerated it away from the other mouth and then back, the mouth going on the journey would age less than the stationary mouth. This is just the twin paradox of the last chapter [169]. Entering through the mouth that was taken on the trip and exiting the other one, one goes backward in time. Oddly enough, this is almost exactly the same situation as posed in *Time for the Stars*; assuming (for lack of any other feasible mechanism) that Tom's and Pat's minds were linked by some sort of flexible wormhole, thoughts going from Tom's mind would go to a decades-younger version of Pat, at least when Tom got close enough to Pat on his return journey to enter Pat's light cone.

There's a snag, however: as mentioned above, Hawking's chronology protection hypothesis states that the universe will not allow time machines to exist. It seems that vacuum fluctuations amplified by the wormhole (probably) destroy it as soon as it becomes a time machine, meaning that once Tom gets close enough to Pat on his return journey, both of their brains are fried by high-energy gamma rays and particles created from the vacuum.

13.7   THE GRANDFATHER PARADOX AND OTHER ODDITIES

> One of the major problems of time travel is not that of accidentally becoming your own father or mother. There is no problem involved in becoming your own father or mother that a broad-minded or well-adjusted family can't cope with.
> —DOUGLAS ADAMS, *THE RESTAURANT AT THE END OF THE UNIVERSE*

A lot has been written about the logical paradoxes involved in time travel. Paul Nahin has examined them in detail in his popular book, *Time Machines* [174]. In chapter 4, "Time Travel Paradoxes and Some of Their Explanations," he covers some of the same ground I cover in this section. I recommend it for anyone interested in the subject. In particular, he discusses John Wheeler and Richard Feynman's reformulation of electrodynamics allowing "advanced" wave (i.e., waves from the future) solutions of the Maxwell equations governing the propagation of light, which I don't cover here [250]. This 1945 work can be seen as a precursor to the work done by Kip Thorne and others on the physical solutions to the "grandfather paradox" discussed below. As the title implies, Paul Nahin's book discusses the philosophical implications of time travel in addition to its physics. This is something I don't cover in detail in this chapter, so his book is a very good complement to the discussion here. He is also a long-time science fiction fan and writer, so the book is written in an engaging style and has copious references to the science fiction literature.

I'm going to write about two of the logical paradoxes in this section. Together, these two probably cover about 99% of what one might encounter either in science fiction or in scientific works. They are:

1. Paradoxes involving creation of matter, energy, or information out of nothing, and
2. Paradoxes involving causality (usually called "grandfather paradoxes").

Paradoxes involving matter, energy, or information creation have to do with the issue that if someone gets into a time machine now and travels back to, say, the Cretaceous period, then we effectively see a large amount of mass-energy disappear *now*. This is forbidden by the most fundamental of the conservation laws of physics, the conservation of mass-energy. In addition, 150 million years ago, a dinosaur observer saw the creation of mass ex nihilo. From then until now, there was extra matter around that wasn't present at the Big Bang. It looks like we got something from nothing.

However, this is easily handled. In the last section I mentioned that local conservation laws imply that a 10,000 kg spacecraft entering one mouth of a wormhole will increase the measured mass by 10,000 kg. The mouth it exits from will lose the same amount of mass. This is justified by calculation of the "back reaction" of the gravitational field as the spaceship passes through it using the general theory of relativity. The only plausible way anyone knows to create a time machine is by using the mouths of a wormhole, so the law of conservation of mass-energy seems to be safe. The mouth through which the time traveler enters gains, and the mouth through which she exits loses, exactly enough to satisfy the law of conservation of mass-energy. Other conservation laws are satisfied as well, such as the conservation of charge. It shouldn't be surprising that the same rules apply, as time travel into the past *is* faster than light travel: in some reference frame, entering the wormhole will happen at the same time as exiting it but separated by a very large distance—nearly 150 million light-years. We can postulate that if time machines or FTL travel exists, both must satisfy the local conservation laws of physics.

Information creation is more tricky. This is a topic many science fiction writers have used in their stories. Let's say I go to the Victoria and Albert Museum in London. In a dim, rarely visited section, I find a paper in a hidden corner that provides detailed instructions for building a time machine. I build the machine and visit H. G. Wells in 1892, telling him how to build the machine. He writes down the instructions, which he hides in a dim, rarely visited section of the V&A. Who figured out how to build the time machine? Where did that information come from?

In examining this paradox we have to look at information in the same way a physicist does to make sense of it. By this I mean let's remove the human factor. Let's imagine we have a wormhole that we have made into a time machine. Going into mouth A causes you to emerge from mouth B at a fixed time interval before entering A. It doesn't have to be long for our purposes. For the sake of definiteness, let's make it 1 ms ($=10^{-3}$ s). Put a computer between the two mouths and run a cable from some output of the computer through mouth A, out mouth B, and into an input into the computer. We now have a computer that sends information to itself in its past.

We write a simple program for the computer to take the input from the computer, calculate some function $f(x)$, and send that to the output. But the output is fed back to the input in the past. Because of this, the input to the program is the output from the program, or

$$x = f(x). \tag{13.12}$$

In mathematical terms, the program must output a fixed point of the function $f(x)$.

Has information been created here? This is a tricky issue, especially if the function's fixed points are hard to determine. For example, if we use the function

$$f(x) = \frac{a}{x}, \tag{13.13}$$

the input/output x will be equal to $\sqrt{a}$, so we have computed a square root using much less computational power than one would have to use normally. This is a trivial example of the computational power available as a result of this "fixed-point" behavior of the computer. Todd Brun has shown that a computer plus time machine can be used to factor large numbers in effectively zero time [42]. This is interesting because the RSA algorithm used to encrypt information is good only so long as factorization remains a "difficult" problem. (Of course, building a time machine represents a difficult problem by itself.) I think that by using this fixed-point principle, one can make Brun's factorization algorithm much simpler by merely using the function call:

$$f(k) = k + MOD(N, k) \tag{13.14}$$

to find out if N has any factors.

One other issue that pops up is that by the appropriate choice of function we can turn this problem into the "grandfather paradox." The grandfather paradox is *the* classic causality paradox. Say your grandfather was an evil man, the dictator of a large country. In his life he was responsible for the deaths of thousands—nay, millions—of innocent people. As his successor, you gather the best scientists in your country together to create a time machine. You go back in time (long before your father was conceived) and shoot the old man to death. So, if your grandfather never fathered your father, who fathered you? Who built the time machine to stop his actions? Time machines seemingly allow the violation of cause-and-effect relations and lead to situations in which we have, paradoxically, two mutually impossible things happening at once.[4]

Again, let's take the human side out of the equation, so to speak, and set up a simple program on our computer: if it gets an input of "1", it outputs a "0", and vice versa. What will it do if we hook up input to output in the way we did above? It's pretty easy to see that there doesn't seem to be any way to satisfy this problem: if a 0 comes out of the wormhole, we send a 1 into it, which sends a 1 back in time as the input to the computer, which then sends a 0 as the output back in time to become the input, which. . . . We can make the "computer" very simple indeed: below is an electric circuit that will mimic what our program will do. The circuit shown is a NOT gate. A 0 V signal (representing a 0 input) produces a 1 V output (standing for a 1 output), and vice versa. The wire from the NOT gate goes through the wormhole mouth A, out B, and back to its input. What happens?

I think there are two possible resolutions to this paradox:

1. Time machines are impossible. If we can't build one, we certainly can't set this apparatus up.
2. Time machines are possible but a paradox is avoided because of the physical limitations of our device.

The first possibility is most likely the correct one but is less interesting, so let's examine the second one. I'll offer an analogy: if you flip a coin, there aren't merely two states possible, heads or tails, but three: the coin can also land on its edge. We think of digital logic circuits as having only two possible states, 0 and 1, but that is an approximation. In reality

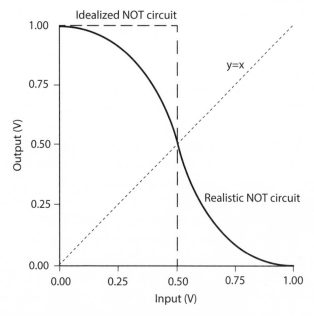

Figure 13.2. Idealized and realistic NOT circuit response curves.

these circuits are built of transistors, devices that obey the laws of physics and that can take a continuous range of input voltages and output a continuous range of voltages. Computer devices use feedback techniques to force them to either go high or low, but these feedbacks fail in the presence of a time machine. My belief, for what it is worth, is that in this situation our computer would find itself in a state that was neither a 1 or a 0 but somewhere between the two. Figure 13.2 shows a realistic response curve for a NOT circuit. Ideally, we would want any voltage under 0.5 V to give us a 1 V output, and any voltage over that to give us a 0 V output. Because the output voltage is a continuous function of the input voltage, the real response will be "softer" than our ideal. As the graph shows, there will be a point where the curve $y = x$ intersects the response curve for the circuit; that is the fixed point. That is what the circuit will output if we hook it up through a time machine.[5]

From this discussion it shouldn't be surprising that time travel into the past, in addition to all of its other problems, leads to violations of the increase in entropy. Computers hooked through time machines can factorize huge numbers simply because we can write programs so that the only consistent output is one of the factors. We can force exceedingly low-probability events to happen because of travel into the past. This is used in a lot of science fiction and fantasy. One can view Rowling's novel *Harry Potter and the Prisoner of Azkaban* in this light. Hermione's time travel forces a lot of low-probability events to occur, including the escape of Buckbeak and Sirius Black, Harry's escape from the Dementors, and so on [202]. In Matt Visser's words, "In the presence of closed timelike curves the consistency conjecture forces certain low probability events to become virtual certainties" [242, p. 256]. Or as Larry Niven wrote, much earlier and more elegantly, "try to save Jesus with a submachine gun, and the gun will *positively* jam" [178].

Finally, I need to address whether the quantum mechanical collapse of a wave function transmits information faster than light. This idea is used in science fiction stories where authors want a quasi-scientific justification for faster-than-light communication such as Ursula K. Le Guin's ansible communicator in *The Left Hand of Darkness* and other novels, and in Daniel Simmon's novels *Hyperion* and *The Fall of Hyperion* [145] [219][220]. In quantum mechanics the wave function replaces the idea of the trajectory of an atom. The wave function is defined at all points in space and time. Its square gives the probability of various properties of the particle: its position, energy, and spin (quantized angular momentum), to name a few. According to the Copenhagen interpretation of quantum mechanics, which most physicists think is correct, these properties don't have values until they are measured [102]. This is a little weird, but not so bad. The difficulty comes when we have two separate particles whose properties are linked together by one or more of the conservation laws. For example, we can produce photons (particles of light) whose spin is anticorrelated because of the law of conservation of angular momentum. In physics terms, if one has spin "up," the other will automatically have spin "down."

Let's do an experiment where we generate these two photons and send them in opposite directions to two observers, Al and Bert, located 2 light-seconds apart. (The experimental apparatus is midway between them.) Al measures his photon as spin "up." The paradox comes in that he knows that Bert will measure his photon to have spin "down" a full 2 seconds before the information from Bert can be transmitted to him. It's even worse when we remember that in some reference frames Al is measuring his photon's spin before Bert measures his. If quantum mechanics is to be believed, the spin of Bert's photon doesn't have any value before it is measured! So what's going on?

This seeming paradox was first discussed by Albert Einstein, Boris Podolsky, and Nathan Rosen in one of the most cited papers in the history of physics, "Can Quantum-Mechanical Description of Reality Be Considered Complete?" [78]. Einstein disliked the conventional probabilistic interpretation of the quantum wave function. The paper was an argument for what are today known as "hidden-variable theories," ones in which the probabilistic interpretation hides a more complicated machinery. The hidden machinery produces well-defined nonprobabilistic values for the quantum mechanical properties of a particle. The EPR paper, as it came to be known, represented a philosophical curiosity for several decades, as there didn't seem to be any good way to test whether the Copenhagen interpretation or hidden-variable theory was correct. This changed in 1964 when John Bell published his paper "On the Einstein-Podolsky-Rosen Paradox" [36]. He showed that certain measurements could distinguish between the conventional view and the hidden-variable theory. All at once the EPR paradox moved to a central position in physics, as there was now a way forward to experimentally test it. Unfortunately, the means to resolve the paradox are complicated, and go beyond the scope of this book. The first experiments were performed in the early 1980s, and the conventional view of quantum mechanics passed with flying colors! Hidden-variable theory simply doesn't work.

We are left with the uncomfortable feeling that the wave function of correlated particles "collapses" faster than light. Because of relativity, it collapses instantaneously in some reference frames and in reversed time order in others. However, there doesn't seem to be any way to transmit information using it. The issue comes down to how we define "information." Because the spin is either up or down with equal probability, Al measures a sequence of random bits on repeated runs of the experiment. He knows that if Bert performs the same measurements, Bert will measure exactly the opposite of what he measures. However, he has no way to know that Bert actually *made* these measurements until Bert transmits the results of his experiments to Al. This can only be done at the speed of light or slower.

These results are interesting for two reasons. First, it made physicists define what "information" meant in a stringent way. Even though Al knows the results of Bert's experiment if Bert makes the measurements, he has no way to tell if Bert made them. If he converts the measurements into a binary code (spin up $= 1$, down $= 0$), all he has is a random string of ones and zeros, which by definition carries no information. Second, in recent years physicists have found applications for Bell's work. There are two main applications of such quantum-correlated measurements. One is in creating secure cryptography. Random strings of ones and zeros are ideal one-time pads for encrypting information. Quantum mechanics guarantees that such strings created by correlated measurements are both perfectly random and uninterceptible [166]. This is good, because quantum correlations are also used in quantum computing. Quantum computing is based on the fact that quantum systems are inherently massively parallel in the language of computer science. Because a particle isn't in any quantum state until it is measured, it is effectively in all quantum states accessible to it. If we act on the particle by different combinations of physical perturbations without measuring the state, it takes all possible paths available to it. It only takes on a settled value once we measure its state. Entangled states, like Al's and Bert's photons, also play a key role in creating quantum computation, although the role that entanglement plays is complicated. If we could build a quantum computer, in principle it could be used to break public key encryption systems.

Public key encryption is used today to secure information for banks, credit cards, and a large section of the global economy. It is based on the fact that it is much easier to multiply two very large primes than factor the composite number back into them using conventional computers. However, if an effective quantum computer can be built, Shor's algorithm could be used to factor them, and all of this information becomes vulnerable. To date, the largest number factored by a quantum processor is 21, but it's a start. All work today on quantum information and quantum computation was motivated by the philosophical and mathematical attempt to reconcile the ideas of relativity theory with quantum mechanics, a process that is far from over.

I'm ending the section of this book on space travel in the same place we began, with a discussion of the physics of computation. Ideas concerning computation have moved from a peripheral position in physics to its very center, as physicists realized that information is a physical quantity like energy and momentum. It is fascinating to see how these ideas have become intertwined with science fiction ideas like the manned exploration of space and faster-than-light travel. Science can be much weirder than science fiction at times.

NOTES

1. Quantum entanglement has also been used as a justification for extra-sensory perception, clairvoyance, and other psionic powers. The word "quantum" is often used to justify anything mysterious or improbable in a story.

2. A tesseract is a four-dimensional hypercube; L'Engle may have gotten the idea of the tesseract as folded space from Heinlein's short story "... And He Built a Crooked House," in which an architect builds a house in the shape of a three-dimensional projection of a tesseract, which then folds itself through four dimensions because of an earthquake. I *really* wanted to live there when I was eleven.

3. Kinda, sorta. . . . It's not a great way to think about it. There's an interesting paper about what things would look like as seen through a wormhole as a result of the strange behavior of light rays entering it: "Visual Appearance of a Morris-Thorne Wormhole," by Thomas Muller [171].

4. Sarah Vowell invented the "grandfather paradox paradox," which investigates the moral paradox of using violence to stop his violence [244].

5. This is a relatively conservative view of what a time machine will do. I have explicitly used the idea that causality isn't violated by time machines. For a discussion of alternative ideas, see Matt Visser's book, *Lorentzian Wormholes* [242].

PART III

WORLDS AND ALIENS

CHAPTER FOURTEEN

DESIGNING A HABITABLE PLANET

> Far too many stories merely give us a planet exactly like Earth except for
> having neither geography nor history.... The process of designing a world
> serves up innumerable story points.
>   —POUL ANDERSON, "HOW TO BUILD A PLANET"

A large number of science fiction stories take place on alien worlds, and
the process of designing an alien world with alien life on it is perhaps
as old as science fiction itself. Hard science fiction writers tend to pay
attention to astrophysics and planetary science to make their worlds
both realistic and exotic, a very difficult combination to achieve. Several
examples of successful designs spring to mind, including Hal Clement's
Mesklin in the novel *Mission of Gravity*, a large planet whose rapid
rotation spun it into an ellipsoid, with a surface gravity three times that of
Earth at its poles but several hundred times Earth's at its equators. There
is also Plateau in Larry Niven's *A Gift from Earth*, a Venus-like planet
with one inhabitable point on it, Mt. Lookitthat, a tall mountain sticking
out of most of the atmosphere [59, 175]. The balancing act one must
go through to create worlds both credible and incredible simultaneously
takes a good deal of work on the writer's part, but is also part of the fun
of the narrative.

The discussion in this chapter is strongly motivated by the wonderful essay "How to Build a Planet," by Poul Anderson. The essay was originally published in the Science Fiction Writers of America bulletin in 1966 and later expanded into one of the famous Writers Chapbooks published by Pulphouse Press. My own copy of the essay is a badly worn Xerox copy of unknown origin. I don't remember when I originally read the essay, only that I was about ten years old and that it was my first exposure to the scientific ideas in science fiction. If I hadn't read the essay I wouldn't have written this book. By necessity, this chapter and that essay cover much the same ground, though in different ways. I would recommend that readers who enjoy this book try to track down a copy of the essay.

14.1   ADLER'S MANTRA

I'm going to stick to a discussion of Earth-like worlds in this chapter. That means worlds that are capable of supporting life as it exists on Earth, sometimes referred to as "carbon-based life." Speculations about non-carbon-based life abound, but since I am a physicist rather than a biologist or chemist, I plan to stick closer to what I know and review the physics of life as we know it. This is still a big issue, but more manageable than if we throw the subject open entirely.

Life as we know it on Earth requires two things at a bare minimum: an atmosphere with large quantities of oxygen in it and average planetary temperatures between the freezing and boiling points of water. The requirements for intelligent life are more restrictive, but we'll deal with that later.

To start with, these two basic requirements probably dictate the type of planet we need to deal with. In our own Solar System there are two basic types of planets, terrestrial and gas planets.

- The terrestrial or rocky planets are Mercury, Venus, Earth, and Mars. Some astronomers also include Earth's Moon as a member of this class of planets. They are the small planets closest to the Sun, and are characterized by compositions that are mostly metals and rocks, with thin to nonexistent atmospheres. The zone of these planets more or less extends out to the asteroid belt, about 2–3 AU from the Sun, and is defined by the so-called "frost line," the distance from the Sun where ices of various kinds form. Until recently it was thought that most solar systems would have the terrestrial planets closest to their sun because of this: as the Solar System was forming, heat from the Sun drove the volatile ices to the outer edge of the Solar System, where they formed the gas giants.

- The gas giants are the four planets of the outer system, Jupiter, Saturn, Uranus, and Neptune. In general, they are characterized by their large sizes compared to the terrestrial planets, with masses ranging from 18 to 318 times the mass of Earth. They are also composed mostly of ice and liquids, with perhaps no solid surfaces.

Other bodies, too small to be called planets, also occur in our Solar System, including dwarf planets, such as Pluto, and over fifty moons, mostly circling the gas giants. Several of these moons might have conditions conducive to life. This has served as the basis of several science fiction stories, the most famous being *2010: Odyssey Two* by Arthur C. Clarke, which centers on the discovery of life on Jupiter's moon Europa [57]. The movie *Avatar* is also set on the habitable moon of a gas giant planet, but not one in our solar system. So is the rebel base in the movie *Star Wars: A New Hope*.

Until 1993, astronomers thought that the arrangement in our Solar System was typical. However, once scientists began discovering planets circling other stars, a third class of planets was discovered, called *hot Jupiters*, planets the size of gas giants but circling their stars at very close distances, sometimes so close that their orbital period is merely hours long! About 25% of all exoplanets discovered are hot Jupiters. To some extent this reflects instrumentation issues: it's easier to discover large planets close to their stars. However, I think it safe to say that no astronomers would have predicted any of these odd giants before they were discovered.

This leads me to a mantra I tell all my astronomy students the very first day of class:

All stars are fundamentally the same; all planets are different from each other.

This doesn't mean that all stars have exactly the same properties or behaviors, but all of the properties of a star stem from two basic data, the star's mass and its composition at the time of its formation. This is known as the *Russell-Vogt theorem* in astrophysics. However, stars are almost identical in their initial composition (mostly hydrogen with a little helium and even less of everything else), so the big determinant of how stars behave, their luminosity, their surface temperature, and their lifetimes is the stellar mass.

On the other hand, planets are a chaotic mess. Although we can put the planets in our Solar System into two broad classes, individual differences among them are as great as their overall similarities. For example, Earth is unique among the terrestrial planets in having a large moon circling it. It is also the only known planet with plate tectonics. There are reasons to think that both these properties might be needed for life on Earth to thrive. Venus is very similar to Earth in size and overall composition, but it has a thick atmosphere with a runaway greenhouse effect, probably a result of being slightly closer to the Sun than Earth is. Mars has a long trench similar to the Grand Canyon but 3,000 miles long, and boasts the largest mountain in the Solar System, the extinct volcano Mons Olympus. Mercury has odd, extremely long cracks running along its surface called lobate scarps, which no other planet has. And so on. The differences result from the fact that many causes determine the characteristics of the planets, and it is not always possible to cleanly separate cause from characteristic.

Here's a short list of causes of planetary characteristics:

- Type of star the planet circles;
- Mean distance of the planet from the star and orbital eccentricity;
- Planetary mass;
- Planetary atmosphere;
- Exact planetary composition; and
- A history of impacts with other objects in the system (especially during formation).

In particular, the history is important. When Poul Anderson mentioned history in his essay, he was probably thinking of the history of the societies on these worlds. In a larger sense, the exact geological history of the planet, and especially the history of planetary impacts, is very important in determining the later features of the planet. If Earth hadn't undergone an impact with a Mars-size object in exactly the right way during its formation, we wouldn't have the Moon. Without the Moon it is very possible that life wouldn't have developed on Earth.

14.2   TYPE OF STAR

Stars are simple objects: all of their properties are determined by mass and composition. The obvious question that arises is, what are those properties? There are really only four important ones:

- Luminosity. Luminosity is how bright the star is and is usually measured relative to our Sun. Luminosities range from about 1/1,000 (10^{-3}) of our Sun's to about 1 million (10^6) times greater than it.
- The surface temperature of the star. The surface temperature runs from about 3,000 K to 30,000 K; our Sun is right in the middle, with a surface temperature of about 5,800 K. Surface temperature also determines the overall color of the star, which ranges from red for the cooler stars to blue-white for the hottest.
- Main-sequence lifetime. About 90% of the stars in the sky burn hydrogen via fusion in their cores into helium. These are referred to as *main-sequence* stars. Our sun is a main-sequence star, and most stories feature planets orbiting main-sequence stars because post-main-sequence evolution probably ends up destroying life on the planets. Main-sequence lifetime is strongly determined by mass: bigger stars live shorter lives.
- Old age and death. After the star eats up all the hydrogen in its core and leaves the main sequence, it evolves into a giant phase, followed by one of three possibilities. For stars with less than about eight times the mass of the sun, the star shrinks and becomes a white dwarf, while for stars between eight and twenty solar masses, the star supernovas, leaving a neutron star behind. The largest stars collapse into black holes.

Stellar modeling is essentially a solved subject. From the 1960s on astrophysicists have developed elaborate computer codes to model the stars,

Table 14.1
Stellar Properties and Spectral Class

Class	T (K)	L (Sun = 1)	M (Sun = 1)	τ (10^9 yr)	R (Sun=1)
O5	42000	5×10^5	60	0.0012	13
B0	30000	32500	17.5	0.0054	7
A0	9800	39.4	2.9	0.74	2
F0	7300	5.21	1.6	3	1.4
G0	5940	1.25	1.05	8	1.1
G2 (Sun)	5800	1	1	10	1
K0	5150	0.552	0.79	14	0.93
M0	3840	0.077	0.51	66	0.63
M7	2860	0.0025	0.08	320	0.2

Source: Data from Carroll and Ostlie [47, appendix G].

and these models have been extensively compared against observation. If any of my readers are interested in the subject I highly recommend the hefty textbook *An Introduction to Modern Astrophysics* by Bradley W. Carroll and Dale A. Ostlie; much of the information presented here is adapted from that book [47]. Main-sequence stars are divided into spectral classes. For historical reasons the classes are (listed from hottest stars to coolest): O, B, A, F, G, K, M, L, and T. Ignoring the two final classes, they can be remembered using the mnemonic "Oh, Be A Fine Girl/Guy, Kiss Me." The classes run in that order from the hottest, brightest, most massive stars to the dimmest, coolest, and smallest. L and T are *brown dwarf* stars, which radiate light mainly in the infrared region of the spectrum and use different fusion processes in their interior; I'll ignore them, although some science fiction stories have been set on worlds circling these stellar objects.

The representative properties of main-sequence stars are a useful way to start thinking about stars for science fiction stories. (The data in table 14.1 and referenced in the following discussion are from Carroll and Ostlie's *An Introduction to Modern Astrophysics* [47, appendix G]).

L is the stellar luminosity (the rate at which the star emits energy in the form of light) in units where the Sun's luminosity is 1; M and R are the mass and radius of the star, again measured with respect to the Sun. The variable τ is the main-sequence lifetime, the amount of time it spends as a middle-aged star, burning hydrogen in its core until all the hydrogen is used up. The numbers after each class represent different subclasses: the lower the subclass: the hotter and brighter the star is.

For any star, the following formula is very useful:

$$L = R^2 \left(\frac{T}{5,800} \right)^4. \tag{14.1}$$

This is the Stefan-Boltzmann formula for a black body emitter in normalized units. We can turn it around and write the temperature in terms of radius and luminosity:

$$T = 5,800\,\text{K} \times \left(\frac{L}{R^2} \right)^{1/4},$$

or write the radius in terms of temperature and luminosity:

$$R = \sqrt{L} \times \left(\frac{5800\,\text{K}}{T} \right)^2.$$

The star's light is emitted in a spectral range that depends on its surface temperature: this is described by the Wien displacement formula from chapter 1, which I'll write in the following way:

$$\lambda_p = 1\,\mu\text{m} \times \frac{2,900}{T}, \tag{14.2}$$

where T is (as always) measured in Kelvin. The reason this is useful is as follows. The smallest stars have surface temperatures right around 2,900 K, so they will emit most of their light in the infrared region of the spectrum (i.e., with wavelengths around $1\mu\text{m}$, longer than what the eye can see). Most of the visible light they emit is in the red end of the spectrum, making them red in appearance. Our Sun, however, has a temperature of 5,800 K, just about twice this, meaning that the light is emitted around a peak wavelength of $0.5\,\mu\text{m}$, right in the middle of the visible spectrum. The brightest stars have temperatures above 29,000 K, so their light will be concentrated at wavelengths of about $0.1\,\mu\text{m}$ or shorter—that is, in the ultraviolet. Most of the visible light they emit will be at short wavelengths, making them appear blue or blue-white.

For main-sequence stars only, the following relation holds:

$$L \approx M^3. \tag{14.3}$$

That is, luminosity increases rapidly with mass, essentially increasing as mass cubed. The power law is an approximation to the true behavior; for very low-mass stars the exponent is somewhat lower, for very high-mass stars the exponent is higher. The fact that luminosity increases so rapidly is why large stars have short lifetimes: a star's main-sequence lifetime is determined by the ratio of the amount of fuel it has to burn (its mass) and the rate at which it is burning it (its luminosity):

$$\tau = 10^{10} \text{ years} \times \frac{M}{L} \approx 10^{10} \text{ years} \times M^{-2}. \tag{14.4}$$

This is an approximate formula. In particular, it doesn't work well for very low-mass stars. One other thing can be worked out: in these units, along the main sequence,

$$R \approx M. \tag{14.5}$$

This isn't a coincidence. It stems from the fact that the proton-proton cycle operating at the core of the star doesn't "turn on" until the core reaches a temperature of about 10^7 K [170].

To convert everything to metric units we need only remember a few numbers. The sun radiates energy at a rate of 3.84×10^{26} W, so

$$L_{metric} = 3.84 \times 10^{26} \text{ W} \times L.$$

Similarly,

$$R_{metric} = 6.95 \times 10^8 \text{ m} \times R,$$

and

$$M_{metric} = 1.99 \times 10^{30} \text{ kg} \times M.$$

With these formulas we can do a lot. In particular, life emerged on Earth only about 700 million years after the Earth formed. As a wild guess, if we assume that it takes life everywhere in the universe about that long to evolve on a given Earth-like planet, we need to pick a star whose main-sequence lifetime will be longer than about 700 million years, meaning that the O- and B-class stars won't work, and possibly not A-class stars either. This doesn't eliminate a lot of them: there are many more smaller, cooler, dimmer, longer-lived stars than there are bigger, brighter, hotter stars. But it gives a point to start from. The smallest class M stars also may not work too well because planets in the zone of life will have to orbit so closely that tidal effects will lock one side of the planet into permanently facing the star, in the same way that the Moon presents only one face to the Earth, making one side much hotter than the other.

Other fun formulas: if the planet is located d astronomical units away from the star, the angular size of the star as seen from the planet is

$$\theta_s = 0.5° \times \frac{R}{d}.$$

If the planet has a moon, its angular size as seen from the planet is

$$\theta_m = 0.5° \times \frac{R_m}{d_m},$$

where R_m is the radius of the planet's moon relative to the radius of Earth's Moon ($= 1{,}737$ km) and a_m is the average distance of the moon from the planet, again relative to the distance of Earth's Moon from the center of Earth ($384{,}000$ km). Kepler's third law tells us that the length of a year on the planet (i.e., the rotational period around the star) is

$$Y = \left(\frac{a^3}{M}\right)^{1/2}. \tag{14.6}$$

where Y is measured in years.

14.3   PLANETARY DISTANCE FROM ITS STAR

Now that we have the star, where do we put the planet? Carbon-based life requires oxygen in the atmosphere and liquid water on the planet's surface. If the planet is too close to its star, it gets too hot, and water will boil on its surface; too far, and water will freeze. In the Solar System there are only three planetary candidates for Earth-like life based on this criterion, Earth, Venus, and Mars.

A planet both absorbs energy from its star and radiates energy on its own. This fact can be used to figure out the temperature of the planet. There are a few ideas that go into this:

1. Radiation from a star spreads out in all directions; the total amount of light the star radiates away is spread out over a sphere centered on the star. If a planet is at distance r away from its star, the total amount of power emitted by the star is spread over a sphere whose surface area is $4\pi r^2$.

2. Some of the light from its star will be absorbed by the planet and some will be reflected away, because of the planetary makeup and the atmosphere of the planet. The total amount of light absorbed will be proportional to the area of the planet.

3. The key point is that the total power absorbed by the planet must equal the total power radiated away (on average); if this weren't true, the planet wouldn't maintain an even temperature. Planetary temperature stays relatively constant because if it radiated away less light than it absorbed, its temperature would increase until it radiated away exactly as much as it absorbed. If it radiated away more light, the temperature would decrease until this happened.

We also have to define a concept called the average *albedo* of the planet. The albedo is the fraction of sunlight that is reflected without being absorbed. Earth's albedo is about 0.3, meaning that about 30% of the light from the Sun isn't absorbed by Earth's surface. This is an average between the oceans (which are relatively absorptive) and the ground and cloud cover (which are relatively reflective), and also changes from point to point and from time to time as well.

Our definition is one that is averaged over the surface of the Earth and over time—say, over several years. I'm also assuming that the orbit is essentially circular. With these definitions, for a planet without an atmosphere,

$$T_0 = 278 \, \text{K} \times \left(\frac{(1-a)L}{d^2} \right)^{1/4}.$$

(14.7)

What this means is that there is a zone of life around the star. As a very rough cut, the zone is the region where the mean temperature ranges from 273 K (on the outer edge), that is, where water freezes, to 373 K, where water boils, on the inner edge. The zone doesn't have hard-and-fast edges because planets have different albedos, and atmosphere plays a role as well. Three points are relevant here:

1. The luminosity of most stars isn't very near that of our Sun. The range of luminosities goes from about 10^{-3} to 10^6—a range over a factor of one billion. This means that the distance of the "zones of life" for different stars will have different values;

2. I call the temperature "T_0" because this is the temperature of the planet if it had no atmosphere. Atmosphere plays a major role in determining planetary temperatures; we'll introduce a simple model for the effect of the atmosphere later in the chapter.

3. This is simply the mean temperature for the planet. The actual temperature will vary quite a bit over time and from point to point on the planet's surface.

We can rework this to find the inner and outer edges of the zone of life for a given planetary albedo and stellar luminosity. Let d_i and d_o be the inner and outer edges of the zone for planets (ignoring the effects of atmosphere):

$$d_i = 0.55\sqrt{(1-a)L},$$

$$d_o = 1.03\sqrt{(1-a)L}.$$

Again, these are very rough cuts. Because we didn't include the effects of planetary atmosphere we find that Earth is actually outside the zone, too far from the Sun, in this naive model. However, it serves as a starting point.

The only candidate planets in our solar system potentially within the zone are Venus, Earth, and Mars. In the late 1800s the astronomers Giovanni Schiaparelli and Percival Lowell wondered whether they had detected signs of life on Mars after seeing an intricate network of canals on its surface; Lowell felt the canals could have been the products of an advanced civilization. This work was the inspiration for innumerable works of science fiction, from H. G. Wells's *War of the Worlds* to Robert Heinlein's *Red Planet* and *Stranger in a Strange Land* [108][115][248]. However, almost from the beginning critics pointed out that Mars was probably too cold and too arid for life of this kind. The Nobel laureate chemist Svante Arrhenius was one of the first to go over the scientific evidence and conclude that Mars was an unlikely abode for life. He was right. The "canals" were merely the product of low telescope resolution and eye strain. But what Arrhenius took with one hand he gave back with the other; in his book, *The Destinies of the Stars* (written with Jones Elias Fries), he predicted that the surface of Venus was wet and misty, with conditions similar to Earth's during its Jurassic period [25]. This prediction led to many a science fiction novel set in the jungles of Venus where dinosaur-like monsters hunted humans through the mud. Heinlein's novel *Podkayne of Mars* is perhaps archetypal of these, although S. M. Stirling has written a new alternate-universe series based on this idea. In these novels, Mars and Venus were seeded with life by an unknown alien race. The feel of them is similar to Burroughs' *Barsoom* novels [118][229].

But these ideas disappeared from serious science in the 1960s and 1970s with the advent of unmanned probes that flew by and landed on Mars and Venus. They found Mars a desert with average temperatures around that of Antarctica, which was not very surprising, and Venus a hell whose surface temperatures reached 750 K—hot enough to melt lead! The difference in the temperature of Venus from our ideal model has everything to do with its atmosphere.

14.4   THE GREENHOUSE EFFECT

Certain gases in the atmospheres of planets tend to trap infrared radiation, which is heat radiated away by planetary surfaces; these gases include carbon dioxide (CO_2), which makes up 380 parts per million in Earth's atmosphere and a whopping 98% of the atmosphere of Venus, and methane, which is found in trace amounts in our atmosphere. The most important greenhouse gas, however, is water vapor, which accounts for about 90% of the total for Earth.

The idea behind the green house effect is pretty straightforward: light from the sun is mostly visible light, which the atmosphere is transparent to; about 70% of it reaches the Earth's surface. However, the Earth radiates away light in the infrared region of the spectrum, so most of it is trapped near the surface. We can see this by considering equation (14.2): because the average temperature of the Earth is about 290, most of the light reradiated from the Earth is at wavelengths near 10 μm, in the infrared region of the spectrum. This is strongly absorbed by the atmosphere, and much of it is reradiated back to the ground. In "building planets," as Anderson puts it, it is vital to include somehow the effects of the atmosphere in our calculations. Computer models to calculate the temperature typically divide up the atmosphere into vertical layers with different amounts of infrared absorbance in each layer; they also grid the world into cells based on the terrain the cells overlie. However, one can get some good insight into what is going on using a simple model.

Greenhouse gases act as a blanket, so that heat from the planet takes longer to escape into space, leading to an increase in the average temperature not predicted by equation (14.7). Mars has a very thin atmosphere, much thinner than the atmosphere at the top of Earth's highest mountains, so this greenhouse effect increases its average temperature by only a small amount. Earth has a moderately thick atmosphere, so the greenhouse effect raises its temperature by about 30 K. Venus has a thick atmosphere mostly composed of CO_2; it has a runaway greenhouse effect, which raises its surface temperature by over 500 K.

Let's assume that the atmosphere can be modeled as one layer, not divided up either horizontally or vertically. We can use a simple model developed in Daniel J. Jacob's book *Atmospheric Chemistry* to calculate the effect of greenhouse warming on Earth. [129, pp. 128–131]. We assume that the planet has an effective albedo a in the visible region of the spectrum and that the atmosphere as a whole traps a fraction f of the light emitted by the surface in the infrared part of the spectrum. Using this model, the temperature of the surface is given by

$$T = 278 \text{ K} \times \left(\frac{(1-a)L}{(1-f/2)d^2} \right)^{1/4}. \tag{14.8}$$

As f goes up, T goes up: the more radiation that is trapped by the atmosphere, the warmer the planet will be. Unfortunately, this is as far as we can go using a simple model; predicting f from the atmospheric constituents is a very difficult task. For Earth's mean temperature of 288 K we need a value $f = 0.77$ with this model. One can use that as a guide when writing science fiction stories set on other worlds.

The temperature of the atmospheric layer can also be found in this model:

$$T_{atm} = T/\sqrt[4]{2} = 0.84 \ T.$$

The model is OK for planets with relatively thin atmospheres, like Earth, but breaks down completely for planets with very absorptive atmospheres, like Venus, where we have to use a multilayer model to come close to the truth. The model also ignores convection and the horizontal transport of energy, both important effects when calculating the real temperature.

For the value of f given above, the mean temperature of Earth will remain between freezing and boiling anywhere within 0.6 to 1.1 AU from the sun. This is not to say, however, that the *real* zone of life is this wide. While planetary temperature depends on atmospheric composition, atmospheric composition also depends on planetary temperature. Venus in many respects is a sister planet to Earth, having about 85% of Earth's mass and being about 90% of its size, but the surface conditions rival hell's. Most astronomers think that Venus became the way it is because of a positive feedback effect. Venus started out with a slightly higher insolation than Earth's, which caused more CO_2 to be liberated from the surface of the planet. This led to more warming, leading to more CO_2

being liberated, and so on [134]. Earth, being about three-tenths of an astronomical unit farther from the Sun, escaped this fate.

Habitable zones also change over time because of changes in planetary atmospheres and solar irradiance. Over time, the Sun's luminosity has increased gradually. Shortly after the formation of the Solar System its luminosity was only 70% of what it is now, and it will be about 10% higher than now in about a billion years [47, fig. 11.1]. From this, Kasting, Whitmire, and Reynolds calculated that over the history of the Solar System from 4.5 billion years ago until today, the habitable zone for Earth-like life extended only from 0.95 to 1.15 AU [136].

It is very unlikely that there was ever any Earth-like life on Venus, given conditions there, but Mars is another story. Mars had a thicker atmosphere once, although it didn't keep it. There is very good geological evidence that liquid water once flowed on the surface of Mars, presumably billions of years ago when its atmosphere was thicker and hence the surface temperature was higher. Where the water is now is anyone's guess; it may be locked in permafrost beneath the surface of the planet. However, conditions once may have permitted Earth-like life on Mars's surface, probably at the bacterial level.

Recent science fiction novels have used this idea, notably Greg Bear's novel *Moving Mars* [33]. In the novel, a terraforming project to make Mars habitable for humans leads to a rebirth of the older life forms on Mars, which became dormant when conditions on the planet became too inhospitable.

The Kepler mission has just begun finding evidence of planets in habitable zones. As of December 2011, the Kepler mission had identified 54 planet candidates within habitable zones, with one candidate, Kepler 22-b, potentially Earth-like (about 2.4 times Earth's radius) [1]. The BBC web report is in the form of a press release; one can find a preprint of a paper concerning this planet submitted for publication at the Cornell preprint server [39]. The writers make it very clear that they do not know that the planet is an Earth-like world. The mass could be up to 124 times the mass of the Earth, and it isn't clear that it is a terrestrial/rocky planet at all. If it is, however, they estimated a value of T_0 (as defined above) of 262 K, assuming an albedo of 0.29 (the same as Earth's), and a temperature of about 295 K if it has an atmosphere with the same properties as Earth's, which they also state is very unlikely.

14.5   ORBITAL ECCENTRICITY

The habitable zone is a spherical shell surrounding a given star. Its borders are fuzzy because it depends on planetary albedo and atmosphere. Unfortunately, planetary orbits aren't circular; they are characterized by orbital eccentricities. Even if the average distance from the star, d, is within the zone, a highly eccentric orbit will take a planet out of the zone for large periods of time as it orbits its star.

Let's say that a planet orbits at an average distance d that is right in the middle of the habitable zone. This "average distance" is really the semi major axis of the planetary ellipse. If the orbit has eccentricity e, then the perihelion and aphelion distances are

$$r_p = d(1 - e), \tag{14.9}$$

$$r_a = d(1 + e). \tag{14.10}$$

If the width of the zone is Δz, then the perihelion and aphelion distances should lie within the zone, meaning that

$$r_a - r_p = 2ed \leq \Delta z,$$

or,

$$e \leq \frac{\Delta z}{2d}. \tag{14.11}$$

Earth's orbital eccentricity is only 0.0167, whereas $\Delta z/d$ for Earth is approximately 0.2, meaning that the eccentricity is well within the bounds set by this limit: Earth remains well inside the habitable zone for the entirety of its orbit. Other planets have larger eccentricities: Mars's eccentricity, the largest in the Solar System, is 0.0934. The values for all the planets in the Solar System are relatively small, but eccentricities for exoplanets can be very large, raising the question of whether our Solar System is typical or atypical in this regard. Ursula K. Le Guin set her novel *The Left Hand of Darkness* on the ice world of Gethen; owing to a combination of relatively large distance and high orbital eccentricity, the world was frozen for most of its long rotational period [145]. It's not entirely clear that life could evolve on such a world, although in the story humans have been settled on Gethen from elsewhere.

Exoplanet data show that most exoplanets have average eccentricities higher than those of the planets in the Solar System; it is not clear whether there is a reason for this or whether it is accidental that our system's planets have low eccentricities. I estimate about 10% of all exoplanets found to date have eccentricities below 0.1. This is ignoring data for hot Jupiters, as tidal friction lowers the eccentricity of their orbits.

14.6   PLANETARY SIZE AND ATMOSPHERIC RETENTION

One other requisite for life (as far as we know) is an atmosphere. It's interesting to compare Earth, Venus, and Mars in this respect. Mars has a very thin atmosphere that is mostly CO_2 (94%, with trace amounts of other gases); Earth has a moderate atmosphere that is 74% N_2, 24 % O_2, and 2% trace elements; and Venus has a horribly thick atmosphere that is almost entirely CO_2.

Mars's atmosphere was thicker in the past.[1] This undoubtedly warmed Mars through to an enhanced greenhouse effect, but the planet lost its atmosphere over several billion years. There are lots of mechanisms by which planets can lose their atmospheres over the course of time. Following are just a few:

- Thermal loss. Molecules in a gas at room temperature or above move with average speeds of a few hundred meters per second, which is about an order of magnitude less than the escape velocity for an Earth-sized planet. However, because this speed represents an average, some of the molecules move a lot faster. In the upper atmosphere, where collisions between molecules are few and far between, molecules moving faster than the escape velocity of the planet can be lost from the atmosphere. This is called the *Jeans escape mechanism*, after the astrophysicist who first described it. There are other non thermal loss mechanisms, including the fact that chemical reactions in the upper atmosphere can provide the reactants with enough energy to escape into space. Generally speaking, larger planets lose their atmospheres more slowly through this mechanism than do smaller ones.

- Impacts. Here is where history becomes important. Impacts of planets with large objects (asteroids or comets) can push a lot of the atmosphere into space. Mars's atmosphere may be thin partly because the planet is close to the asteroid belt and suffered collisions from asteroids over geological time spans.

- Solar wind. Stellar winds, streams of charged particles from the sun, can strip away planetary atmospheres, particularly for small worlds close to their stars. This may have affected Mars's atmosphere and almost certainly is responsible for stripping away what little atmosphere Mercury may once have had. In fact, all of the atmospheres of the inner planets are secondary ones, as the primary atmospheres that formed when the planets formed were stripped away by the strong solar winds from the young Sun.

- Chemical sequestration. The atmospheres can become chemically bonded to the crust of the planet. This is where most of Earth's carbon is: if all the carbon in the Earth's crust were liberated into the atmosphere, Earth would have a worse greenhouse effect than Venus.

The formation of atmospheres is just as complicated, and harder to discuss in detail, as it is due to vulcanism, cometary impacts during the planet's early history, and, for Earth, the presence of life on the planet (leading to the presence of significant amounts of O_2 in the atmosphere).

14.6.1 Thermal Loss Mechanisms

The average (rms) speed of a molecule of molar mass m in grams per mole in a gas at temperature T is given by

$$v_{rms} = 157 \, \text{m/s} \times \sqrt{T/m}, \tag{14.12}$$

whereas the escape velocity of a planet is given by

$$v_e = 11,000 \, \text{m/s} \times \sqrt{M_p/R_p}, \tag{14.13}$$

Table 14.2
Relative Properties of Earth and Mars

	Earth	Mars
R	1	0.533
M	1	0.107
T	1,000	140
V_e (m/s)	11,000	4,800
v_{rms} for H_2	3,500	1,860
He	2,480	930
O_2	878	328

where M_p is the mass of the planet measured relative to the mass of the Earth and R_p is the radius, again measured with respect to the Earth's radius. A rule of thumb derivable from the Jeans escape formula that is used by planetary scientists is that if the rms molecular speed is greater than about one-sixth the escape velocity, thermal loss mechanisms will deplete the atmosphere of that particular molecule over geological periods of time [46, p. 103].

Table 14.2 shows some comparisons for Mars and Earth. The temperatures were chosen as typical of the top of each planet's atmosphere. Earth is at a relatively high temperature because of the ozone layer. The absorption of UV light from the Sun puts energy into the atmosphere at that height, while Mars doesn't have a similar protective layer.

There are a few points we can glean from table 14.2:

1. Neither hydrogen nor helium should be present in either atmosphere in appreciable quantities because their average molecular speeds are too high.
2. However, oxygen shouldn't be lost because of thermal effects in either atmosphere. In the case of Mars, the low escape velocity is offset by the low atmospheric temperature compared to Earth's.

The ratio of the average molecular speeds to the escape velocity is just about the same for molecules in the atmospheres of each planet. We therefore cannot attribute differences in atmospheric composition in either planet to purely thermal effects.

There is almost no oxygen in the Martian atmosphere because it is bound up in the Martian soil in the form of Fe_2O_3—rust. Oxygen is so highly reactive that chemical sequestration will bind it unless there is a continual source of it from somewhere else. In the case of Earth, the source of oxygen is the respiration cycle, that is, it is due to life on Earth. Similarly, there is almost no water on Mars because UV light dissociated the water vapor into hydrogen and oxygen; the oxygen was sequestered in the soil, while the hydrogen escaped into space. This shows how complicated atmospheres can be. That Mars has a much more rarified atmosphere than Earth doesn't seem to owe principally to its mass and size but to a number of complicated factors.

14.6.2 Impacts

The total mass of Earth's atmosphere is roughly 4×10^{18} kg. If we wanted to get 1% of the total mass of the atmosphere to escape velocity we would need to supply it with an energy of about 2.5×10^{24} J. Comets or asteroids typically hit Earth at speeds of about 30,000 m/s, that is, at about the orbital speed of Earth. A collision of this energy would require an impact of an object with mass about 5×10^{15} kg, or, assuming an average density of 5,000 kg/m^3, a volume of about 10^{12} m—the equivalent volume of a cube 10 km on a side.

This is a very large impactor, about the size of the comet that wiped out the dinosaurs. Such impacts happen to Earth only about once every hundred million years or so. Because of this, we are pretty safe in ignoring this atmospheric loss mechanism for Earth, at least under present Solar System conditions. Comparing Mars to Earth in this manner is interesting. Mars is less massive than Earth, meaning it has a lower escape velocity, and it is closer to the asteroid belt, meaning it will sustain more frequent impacts. Both factors favor atmospheric loss from impacts for Mars over Earth.

14.6.3 What Is the Range of Sizes for a Habitable Planet?

From all of this it seems that are no easy criteria with which to establish a lower bound on the size of a planet capable of supporting life.

If Mars had an ozone layer, would it have kept its water vapor, leading to higher planetary temperatures from greenhouse warming, or would it have lost its atmosphere faster because it would have developed a thermosphere similar to Earth's? If we put Mars in Earth's orbit, would it have kept an atmosphere longer because it didn't suffer from as many impacts? As a guess, I would say that Mars is close to the lower bound on the mass or radius for a habitable world, as it seems that conditions there are right on the cusp of allowing life. This speculation must be taken with a large grain of salt: because of the interrelation of all of these variables it is hard to give definitive answers.

How about the upper limit on a habitable planet's mass? This is similarly hard to estimate. One upper bound is that if a planet retains lighter gases in its atmosphere, it will likely turn into a gas giant planet. However, this depends both on the planetary mass and on its position in the Solar System, as all the gas giant planets formed beyond the "frost line," outside the orbit of Mars. Perhaps more to the point, all the gas giant planets are thought to have solid cores of about ten times the mass of the Earth; maybe this core mass represents an upper limit. Or maybe not.

14.7   THE *ANNA KARENINA* PRINCIPLE AND HABITABLE PLANETS

All happy families are fundamentally the same; each unhappy family is unhappy in its own way.
—LEO TOLSTOY, *ANNA KARENINA*

Thus reads the famous opening line of Tolstoy's *Anna Karenina* novel. Jared Diamond in *Guns, Germs, and Steel* introduced what he referred to as the "Anna Karenina" principle when reflecting on why, out of all possible animal species on the planet, only a handful had been domesticated by humans. He found that all animals domesticated for food had a number of features in common: they were all herbivores, matured quickly, bred in captivity, and had a few other similarities.

To quote Diamond,

> To be domesticated, a candidate species must possess many different charac-
> teristics. Lack of any single required characteristic dooms efforts at domesti-
> cation, just as it dooms efforts at building a happy marriage.

As he put it, "For most important things... success actually requires avoiding many separate causes of failure" [65, p. 157].

Above I introduced Adler's mantra: "All stars are fundamentally the same; all planets are different from each other." For planets, I will rephrase slightly:

> Each lifeless planet is different from the others in its own way; all planets with
> Earth-like life on them will be fundamentally the same.

This is an extension of the Anna Karenina principle as applied to habitable worlds: all Earth-like planets have a number of similar characteristics, the most obvious being that they fall within a "zone of life," not too far from or too near their star. The realization of this point has solved a conundrum that has faced scientists for a very long time: if the Earth is an average planet circling an average star in an average galaxy, with nothing special about it, then why haven't we found life elsewhere in the cosmos yet? Why isn't the universe teeming with life? Why haven't the aliens made contact?

Since the 1960s, two ideas have gradually developed due to advances in planetary science:

1. The conditions on the different planets are far more diverse than was realized before. Planetary formation seems very chaotic, and planetary history (among other factors) plays a larger role in determining the geological and climatological features of the terrestrial planets than anyone realized.

2. The conditions required for Earth-like life on a planet are far more restrictive than was thought in the 1960s.

These two realizations severely reduce the number of possible planets with Earth-like life. Although Earth is in some sense no more special than any other planet, it is special in other ways as a cosmic lottery winner: it got everything right for life to appear on it. The point is that while it is improbable for any one particular person to win the lottery, someone almost always wins it.

Probability estimates of the number of worlds with life on them, in the fashion of the Drake equation, are meaningless, as we simply don't have enough data. The criteria I have given so far are pretty solid for any planets with Earth-like life, but in the next section I'll list a number of criteria for which there is less solid evidence.

14.8 IMPONDERABLES

One complication concerning Mars has been discovered by numerical simulations of its rotation. One of the characteristics of Earth's climate is its long-term stability. This results in part from the fact that the rotational axis of the Earth more or less points in the same direction for long periods of time and doesn't change dramatically.[2] However, because of Mars's elongated orbit and its lack of a large moon to stabilize it, the orientation of the rotation axis of Mars can change dramatically and chaotically over the course of millions of years. It is believed that this has dramatically changed the characteristics of seasonal change on Mars [135][144]. It isn't clear whether a stable rotational axis is needed for the evolution of life on a planet, but a stable climate certainly helps, in which case having a large moon might be a requirement for planetary life.

Another issue is that Earth is currently the only known planet that experiences plate tectonics, a result of both its size and its composition (radioactive decay in the Earth's core keeps the mantle plastic). Some scientists have speculated that plate movement contributes to evolution because it allows the broad dissemination of plant and animal species, making it harder for individual species to be wiped out by a local catastrophe. Who knows? I am not aware of any science fiction stories that have incorporated the relationship between plate tectonics and evolution thematically; it would be interesting to see if it could be done in any reasonable way.

Other issues: Most exoplanets are found around stars with high metallicities, that is, around stars that contain more metals than average.

(To an astronomer, a metal is any element that is not hydrogen or helium.) It may be that stars with higher metallicity simply contain more of the stuff that planets form from, although the exact details are not entirely clear [100]. More than half of all stars found so far with exoplanets have even higher metallicities than the Sun. What is interesting, however, is that high-metallicity stars are relatively young (population I) stars because the metals were mostly created by higher-order fusion processes in the hearts of the stars; the oldest stars formed at a time when these elements simply didn't exist [130, pp. 495–497]. Because the higher elements are distributed by supernova explosions, which are more common toward the center of the galaxy than in the spiral arms, there may be a higher occurrence of stars with planets toward the centers of galaxies. However, the higher incidence of supernovas and radiation from supernovas may sterilize life emerging on these planets at distances too close to the galactic center. Thus there may also be a "galactic habitable zone" where life may form, an annulus neither too close to nor too far from the center of the galaxy [100].

Giant planets on highly eccentric orbits may perturb Earth-like planets out of the life zone because of their gravitational interactions with them, but large planets in the outer system may serve to screen planets from asteroid impacts like the one that eradicated the dinosaurs [247]. Hot Jupiters are thought to form in the outer system but migrate by various processes to close orbits around the star; the migration may disrupt the formation of planets in the zone. So the presence of hot Jupiters (found in over 10% of all exoplanet systems to date) may preclude life from developing, but cold Jupiters on nearly circular orbits in the outer system may be needed for life.

There is an almost infinite list of considerations which one can go into, especially when we add the question of intelligent life to the mix. I consider this subject in a later chapter. For anyone interested in exploring these ideas further, Brownlee's book *Rare Earth* is a good place to start, but there has been a lot of research in this area since the book was published in 2000 [247].

We have now listed the criteria for a planet to support Earth-like life. In the next chapter I take up the issue of actually finding it out there.

NOTES

1. In a prescient piece of writing, Edgar Rice Burroughs in *A Princess of Mars* wrote that the Martian civilization built an atmosphere plant to combat atmospheric loss from their world. The means by which the atmosphere was replenished (the "ninth ray") are not particularly scientific, however [43].

2. Dante Alighieri in 1300 CE understood this, though in a somewhat different way than we do now.

CHAPTER FIFTEEN

THE SCIENTIFIC SEARCH FOR SPOCK

Heaven and earth are large, yet in the whole empty space they are but as a small grain of rice. . . . It is as if the whole empty space were a tree, and heaven and earth were one of its fruits. Empty space is like a kingdom, and heaven and earth no more than a single individual person in that kingdom. Upon one tree there are many fruits, and in one kingdom many people. How unreasonable it would be to suppose that besides the heaven and earth which we can see there are no other heavens and no other earths.

—TENG MU, *PO-YA CH'IN*

15.1 EXOPLANETS AND EXOPLANTS

The idea of life in other stellar systems is an old one and a source of speculation at least since the times of the ancient Greeks. However, serious scientific attempts to detect life outside the Solar System dates back only to the 1960s, and at that time were a marginal effort. Although a number of well-known scientists including Carl Sagan and Philip

Morrison participated in the search, it was never well funded. It always remained a research sideline even for those people most passionately interested in it.

This changed significantly in the 1990s, when it moved from the sidelines to a central place in modern astronomy. It is now funded at a rate hundreds of times what it was before then. The reasons have much more to do with developing technology than with the amount of interest in the subject.

The search for extraterrestrial life has its roots in the 1800s, when physicists and chemists began to realize that life was a physical and chemical process, not something separate from these subjects. As I mentioned in the last chapter, the American astronomer Percival Lowell claimed to have seen canals on Mars through a telescope and speculated they could have been the product of an advanced technological civilization [155, chapter 4]. In *The Destinies of the Stars,* published about 20 years after Lowell first made his claims, Svante Arrhenius and Joens Elias Fries discussed the limitations to this hypothesis; they pointed out that spectrometers had detected no water vapor in the Martian atmosphere, making it improbable that the canals existed [25, p. 183]. This didn't stop three generations of science fiction writers, from H. G. Wells and Edgar Rice Burroughs to Robert Heinlein and Ray Bradbury, from using the idea of the ancient dying Martian civilizations in their works [40] [43] [108] [118] [248]. What is of note here is that Arrhenius used state-of-the-art technology, photographic spectrograms of Mars, to refute Lowell's argument. Science is often driven by available technology, and nowhere is this more true than in the search for life in the universe. Of course, new technology and new discoveries often raise as many questions as they answer: *Mariner 9* pretty much ended any serious ideas of an advanced Martian civilization, but it did provide evidence (since confirmed by later probes such as the Mars Rovers) that climate conditions in the past were more favorable for life, Mars having gone through warmer periods when water flowed openly on its surface [163] [193] [208]. This suggested the possibility that instead of life in the form of an advanced technological civilization older than humanity, it may have existed on Mars in a more primitive form, at the bacterial level or as simple plant life.

In the 1960s it became clear that there was no advanced life elsewhere in the Solar System, so attention turned to life on hypothetical planets in other stellar systems. However, it is prohibitively expensive in energy costs to travel to other star systems, and, relativity being what it is, it takes a long time, too. Therefore, the focus turned to the idea of detecting life outside the Solar System. Giuseppe Cocconi and Philip Morrison pointed out in 1959 that radio telescopes could detect radio or microwave signals sent by an advanced alien civilization from a distance of more than 10 light years away [60]. Now, the 1960s were *the* era of the radio telescope. It was during this period that radio telescopes detected the background heat of the Big Bang and pulsars, among a host of other discoveries. In fact, in 1962, when Jocelyn Bell Burnell detected radio signals from the first discovered pulsar, the periodic signal was originally thought to be from aliens contacting Earth; the signal was originally given the name LGM-1, standing for "Little Green Man." I call the years from 1959 to 1993 the SETI period because that is when the search for extraterrestrial intelligence flourished. I devote chapter 12 to the subject of the detection of alien intelligence in detail; for now, I will simply mention that several searches for alien intelligence, initially organized by Carl Sagan and Frank Drake and later funded by Steven Spielberg and others, looked for but did not find any alien races attempting to contact us over a radio or microwave channel.

In the 1960s it was impossible to detect exoplanets—planets outside the Solar System—using optical telescopes. This is why scientists wanted to look for intelligent life signaling us; it was the only way we could detect it. But things have changed: in 1993, ground-based optical telescopes detected the first planets circling other stars, and since then over 700 exoplanets have been confirmed, with over 2,000 other candidates; the planets within our Solar System are now in the minority by a huge margin. Most planets have been discovered using space-based instrumentation, first the Hubble Space Telescope and more recently the Kepler mission instruments, but a host of other Earth- and space-based detectors have found them as well.

Even with modern technology, few exoplanets have been directly imaged. The reason is that stars are large and bright, whereas planets are small, dim, and close to the star. If the star is D light-years away and the planet is d astronomical units away from the star, the angular separation between them as seen from Earth is given by

$$\theta = 3.26'' \times \frac{d}{D}. \tag{15.1}$$

One arc-second ($''$) is 1/3,600 of a degree; this is the apparent size of a quarter seen at a distance of fifty football fields. This telescopic resolution is not impossible: the best telescopic angular resolution depends on the wavelength of light and the diameter of the telescope. Most exoplanet searches are done using telescopes designed for work in the near infrared or visible region of the spectrum where the wavelength of light used is near $1\,\mu$m ($= 10^{-6}$ m). If A is the diameter of the aperture of the telescope and λ the wavelength being used by the telescope, in the units given above the smallest angle between two objects that the telescope can possibly resolve is given by the Rayleigh resolution criterion:

$$\theta_{best} = 0.25'' \times \frac{\lambda\,(\mu\text{m})}{A\,(\text{m})}. \tag{15.2}$$

From this criterion a telescope with a 1 m diameter operating at a wavelength of $0.5\,\mu$m (in the middle of the visible spectrum) could resolve this angular separation for a planet 1 AU away from its star at a distance of 26 light-years. Ideal conditions are rarely met on Earth, where atmospheric turbulence reduces telescope resolution considerably, but telescopes in space can operate with near ideal resolution. The Hubble Telescope, for example, has a $0.0403''$ resolution in the visible region of the spectrum ($0.5\,\mu$m) and a $0.026''$ resolution in the ultraviolet region of the spectrum [7].

Even with a telescope with sufficient resolution, a planet is still difficult to find because it is very dim compared to the star it orbits: it will reflect a certain amount of light from the star in the visible region of the spectrum and emit light in the infrared because of its own temperature, but those amounts will be very small compared to the star's. The luminosity of the planet in the visible and the infrared region is given by the formulas

$$L_{p,V} = aL \times \left(\frac{r}{2d}\right)^2 \tag{15.3}$$

and

$$L_{p,I} = (1 - a)L \times \left(\frac{r}{2d}\right)^2, \tag{15.4}$$

where V and I stand for visible and infrared, respectively, L is the luminosity of the star, a is its average albedo in the visible region of the spectrum, d, as above, is the distance of the planet from the star, and r is the radius of the planet. The distance and radius can be in any units you want so long as they are the same units.

For example, the Earth has a radius of 6,500 km and a distance from the Sun of 1 AU, or 1.5×10^8 km; from this, its visible luminosity is only about 1.4×10^{-10} of the Sun's and its infrared luminosity is only 3.4×10^{-10} of the Sun's.[1] Despite these problems, a few exoplanets have been directly imaged by both space-based and ground-based telescopes using various clever techniques. The first exoplanet imaged was a planet orbiting the bright, close star Fomalhaut; the planet, known as Fomalhaut b and discovered in 2008, is about three times the mass of Jupiter and orbits at a distance of 115 AU [5].

Direct imaging works best for large planets far away from their star. Because of the difficulties in seeing such small objects so far away, only twenty-nine planets have been detected by direct imaging out of more than 700 confirmed exoplanet finds and nearly 2,000 more candidates [4]. Most planets outside the Solar System have been detected by indirect means. There are two major indirect methods for finding exoplanets: by Doppler wobble and by transits. The physics behind each of these methods is interesting enough that we should look at them in detail.

15.2 DOPPLER TECHNIQUE

In earlier chapters we discussed Kepler's laws of planetary motions and went over the fact that planets orbit their stars under the influence of gravity on elliptical orbits. According to Newton's third law, this isn't the entire story: for every action there is an equal and opposite reaction. The star itself is influenced by the gravitational attraction of the planet.

In fact, the planet doesn't really orbit the star; both star and the planet orbit a common center of mass (which is located quite close to the star). The orbit of the planet is big, and the planet travels it with a high velocity; the motion of the star is small, and the star orbits the center of mass with a low velocity. *But the star's motion is measurable.*

Let's say that we have a star of mass M with a planet of mass m at distance d from the star. To make things simple, we'll consider just the case of circular orbits: if the star's orbital speed around the center of mass is V and the planet's speed is v, then conservation of momentum tells us that

$$MV = mv, \tag{15.5}$$

or

$$v = \frac{M}{m}V.$$

Since m is much smaller than M, the planet's speed will be much bigger than the star's speed. As an example, Jupiter, the largest planet in our Solar System, has a mass 318 times the mass of the Earth, which is just about 1/1,000 the mass of the Sun. Since its orbital speed is 13 km/s, the orbital speed of the Sun is a mere 13 m/s, and the effects of Jupiter will dwarf the effects of the other planets in the Solar System.

The Doppler technique gives two key pieces of information:

1. The speed of the star as it circles the center of mass, and
2. The period of the planet as it circles its star. The issue here is that using the Doppler effect we can see whether the star is moving away from us or toward us. As the planet circles its star, the star is sometimes moving toward us, sometimes away from us; the time it takes to make one complete period is the same amount of time it takes the planet to circle its star.

From these pieces of information we can get the speed of the planet around the star, the distance of the planet from the star, and the mass of the planet. Here's how it works:

If Y is the amount of time it takes for the planet to circle the star (measured in years), d is the average distance from the star (measured in astronomical units), and M is the mass of the star (measured relative to our Sun), then Kepler's third law tells us

$$\frac{d^3}{Y^2} = M. \tag{15.6}$$

But we know M because (on the main sequence) there is a strong correlation between spectral class and stellar mass (as detailed in the last chapter), and the spectral class of a star is easy to determine. The speed of a planet on a circular orbit around its star is given by the formula

$$v = 30 \, \text{km/s} \times \sqrt{\frac{M}{d}}, \tag{15.7}$$

which can also be expressed as

$$v = 30 \, \text{km/s} \times \frac{d}{Y}.$$

So we know the distance of the planet from the star because we have measured M and Y:

$$d = (MY^2)^{1/3}. \tag{15.8}$$

We also know v:

$$v = 30 \, \text{km/s} \times \frac{d}{Y} = 30 \, \text{km/s} \times \left(\frac{M}{Y}\right)^{1/3}. \tag{15.9}$$

From v we get V, and now we know the mass of the planet. Because Jupiter's mass is almost exactly 1/1,000 the mass of the Sun and 1,000 meters = 1 km, we can write a simple formula for the mass of the exoplanet in terms of Jupiter's mass:

$$m \, (m_J) = M \, (M_s) \times \frac{V \, (\text{m/s})}{v \, (\text{km/s})}. \tag{15.10}$$

Again, to be very clear, to use this formula, express the planet's mass in units where the mass of Jupiter equals 1, express the mass of the star in units where the mass of the Sun equals 1, express the planetary speed in kilometers per second, and express the speed of the star in meters per second.

Actually, this technique really only gives a lower limit to the mass of the planet: you can only use the Doppler effect to measure the motion of the star either toward or away from you, but the plane of the orbit can be at any orientation with respect to the telescope. Because of this, what we are really measuring is $m \sin i$, where i is the inclination of the orbit: $i = 90$ degrees means that the orbital plane is in line with the telescope, while $i = 0$ degrees means that we are looking at it perpendicularly.

15.3   TRANSITS AND THE KEPLER MISSION

The other commonly used technique to detect exoplanets is the transit method. If the planet's orbit is at an inclination near 90 degrees, the planet will pass almost directly in front of the star as seen through a telescope on Earth. When the planet passes in front of the star, the star will dim by a tiny amount. The fractional dimming is equal to

$$f = \left(\frac{r}{R}\right)^2, \tag{15.11}$$

where r is the planet's radius and R is the star's. Since Earth's radius is only 1/100 the radius of the Sun, aliens trying to see the Earth through transits would need to measure a fractional dimming of the sun of $(1/100)^2$, or one part in 10,000. Jupiter, on the other hand, is about 10% the size of the Sun, so we would need to measure a fractional dimming of only one part in 100. This isn't great, but it beats the one part in 10 billion accuracy needed to directly image the planet. One can potentially measure three things from this technique:

- The radius of the planet, which can be calculated from the fractional dimming.
- The period of the planetary orbit. Each time the planet orbits the star, the star is dimmed, so watching the star for a long time tells you how long it takes for the planet to circle the star.

- The orbital inclination, which can be derived from how long it takes the transit to occur and the size of the planet.

You can't get the mass directly from this, but if you can measure the mass using the Doppler technique as well, you can determine the composition of the planet in a crude way because knowing the size and mass allows one to determine an average density. The Kepler mission uses a specially dedicated space telescope to look for such transits; it has found more than 2,300 planet candidates, with 74 confirmed planets.

Both Doppler and transit techniques work better if the planet is large and close to its star. For this reason, the planets found to date tend to be the size of Jupiter or larger, and also closer to the parent star. However, as of December 2011 the Kepler mission had found several planets about the same size as Earth and one "super-Earth" in the habitable zone of its star [6].

15.4   THE SPECTRAL SIGNATURES OF LIFE

Once planets in the zone are detected, can we find life on them from a distance of light-years? The answer depends on whether we can measure the composition of any atmosphere the planet has in a transit. This is a daunting task. Earth's atmosphere, for example, extends only about 100 km above the surface of the planet. If the Earth were shrunk to the size of a basketball, the atmospheric layer would extend only about one-tenth of an inch above it. The atmosphere is mostly transparent; changes in the spectrum of the star owing to the atmosphere will only be a tiny fraction of the changes owing to the transit itself, which is only a tiny change in the star's brightness to begin with. However, these changes have been detected. In 2002 a team led by David Charbonneau measured the presence of sodium in the atmosphere of a planet orbiting the star HD 209458; the planet has a radius about 1.35 times Jupiter's. The fractional change in the spectrum owing to the atmosphere was only about two parts in 10,000 [49].

Since then, astronomers have detected atmospheric constituents on more than a dozen exoplanets [213]. The transit technique works best in detecting atmospheres for hot Jupiters as they are large and close to their star, making the signal relatively large and the probability of a transit high. Atmospheric water vapor, carbon monoxide, carbon dioxide, and methane have all been seen as atmospheric constituents for these types of planets. To date, no atmospheres for terrestrial planets or super-Earths have been detected, but it is only a matter of time. The hope is in the detection of either oxygen or ozone (O_3) for a terrestrial planet in the habitable zone around a star; because oxygen is highly reactive, its presence in an atmosphere is almost a guarantee of the presence of life on the planet. Recent studies indicate that both should be detectable in exoplanet atmospheres [214]. There are other signatures of life as well, however. Some are discussed in the next section.

15.5   ALIEN PHOTOSYNTHESIS

David Gerrold's The War against the Chtorr series is one of the most interesting perspectives on alien invasion novels that there is. In the novels, the aliens aren't invading Earth with a direct display of technological force; instead, there is a wholesale displacement of Earth's ecology by a variety of alien plants and animals, each one worse than the last. The perhaps dominant Chtorran species are large wormlike beings, covered in purple-orange "fur," that eat anything that moves. Apart from Gerrold's tendency to polemicize endlessly, the books are fascinating and show the real potential of science fiction as a literature of the intellect. (If you are reading this book, David, please finish the series before you die!) The books are wonderful in that humanity has no idea whether there is a dominating intelligence behind the invasion that hasn't shown itself yet or whether the alien ecosystem as a whole is invading out of some collective unconscious action.

I want to take exception with Gerrold on one matter: in the first novel in the series, *A Matter for Men*, the hero, Jim McCarthy, experiments with several samples of captured Chtorran life and finds that they are much more responsive to reddish light than to typical daylight:

> "The point is, the atmosphere is thick and the primary is dim, but how much of each I don't know—oh, but I can tell you what color it is."

"Huh?" Jerry's jaw dropped. "How?"

"That's what I've been working on. . . . The star is dark red. What else?"

Jerry considered that. His face was thoughtful. "That's fairly well along the sequence. I can see why the Chtorrans might be looking for a new home; the old one's wearing out." [92, p. 151]

Later on this piece of evidence is used to deduce that the Chtorran ecology may be more advanced than Earth's by half a billion years because of the advanced age of its star [92, p. 211]. I want to dispute Jerry's and Jim's conclusions here: what they have implicitly stated is that the Chtorran star was originally a star like our Sun but is old enough to have evolved off the main sequence and into a red giant. However, this inference is completely unwarranted: when our Sun evolves off the main sequence, it is likely to sterilize life on Earth. The Chtorrans would have had to have left their planet much earlier than that to avoid being destroyed. In fact, life on Earth may only be possible for another billion years before increasing luminosity from the Sun creates a runaway greenhouse effect like that on Venus, long before Sol evolves off the main sequence [136].

It is much more likely that the Chtorran homeworld circles an M-class main-sequence star: such stars are very common, their dominant radiation is in the red or infrared region, and they are extremely long-lived, meaning that life on these worlds could have existed for a much longer time than on Earth. Another interesting point in the book is that Chtorran "vegetation" such as "red kudzu" is, as the name implies, bright red. This raises an interesting question: what color would alien vegetation be?

Vegetation on Earth is green because it dominantly absorbs light in the red region of the spectrum (because there are a lot of photons there) and in the blue region of the spectrum (because they are high energy); the middle region of the visible spectrum, in the green, is reflected dominantly, leading to the plant's colors. However, the chemistry and physics of photosynthesis are very complicated, as no less than eight photons are involved in each photosynthetic reaction [245]. N.Y. Kiang and her colleagues at NASA considered the twin questions of why photosynthesis on Earth evolved to absorb in the spectral region it does and what photosynthesis might be like on a planet circling another star [138] [139]. Kiang also published a less technical version of these two articles in *Scientific American* [137]. To quote from the latter article:

> The range of M-star temperatures makes possible a very wide variation in alien plant colors. A planet around a quiescent M star would receive about half the energy that Earth receives from our sun. Although that is plenty for living things to harvest—about 60 times more than the minimum needed for shade-adapted Earth plants—most of the photons are near-infrared. Evolution might favor a greater variety of photosynthetic pigments to pick out the full range of visible and infrared light. With little light reflected, plants might even look black to our eyes [137].

So red kudzu is definitely possible, but black kudzu might be more likely. One issue with vegetation around M-class stars is that they have frequent UV flares, which have the potential of preventing life from establishing itself on land, confining it to a region below 9 m beneath the surface of the water.

Kiang and her colleagues also considered plant life on planets circling F- and K-class stars. For the F-class stars, their conclusions were that plants might have a predominantly bluish coloration. This would be another spectral signature of life, producing an "edge" in the spectral area where photosynthesis reflected light away. Their conclusion was that it is plausible but not proven that the spectral signature of plant life could be detected over interstellar distances. At the very least, her articles give science fiction writers something new to chew over [139].

The reason why the search for life in the universe nowadays centers on the search for vegetation is very simple: we can find it without it having to signal us. Waiting for the aliens to call was always dicey and hinged on a lot of arguments that a lot of people considered unscientific. However, the possibility of vegetation on other worlds relies on well-understood scientific principles. For this reason the search for alien life is much better funded today than it was forty years ago: it has a much better chance of success. However, finding alien vegetation or bacteria is not nearly as exciting as finding other intelligences—someone to talk to. I take this up in the next chapter.

NOTES

1. One subtlety, however, is that *relatively speaking* its effective luminosity in the infrared compared to the Sun's luminosity *in the infrared* will be higher than the number quoted because the Sun emits most of its light in the visible region of the spectrum. The Sun is a weak emitter at wavelengths near 10 μm, which is where the Earth emits most of its light.

CHAPTER SIXTEEN

THE MATHEMATICS OF TALKING WITH ALIENS

16.1 THREE VIEWS OF ALIEN INTELLIGENCES

Apart from space travel, contact with alien races and civilizations is *the* theme of science fiction. In this chapter we will examine the issue of communication with alien races. There are, of course, many hurdles to overcome in any communication with intelligent aliens. The first is whether any exist at all. Science fiction writers tend to adopt one of three views:

1. Alien life is common across the galaxy.
2. Alien life is rare, but intelligent aliens exist.
3. Alien life is nonexistent.

Let's discuss each of these views in turn.

16.1.1 Alien Life Is Common across the Galaxy

The view that alien life is common across the galaxy is perhaps the most typical view in science fiction stories written from the early 1900s to the

1980s. In popular culture, the TV show *Star Trek* in the 1960s exemplified this view. The show featured a new alien culture and civilization almost every week, with recurring examples of the Vulcans, Klingons, and Romulans. Each of the new series in the franchise introduced new intelligent alien species until there are probably now several hundred in the *Star Trek* canon. To some extent this is because many of the writers for the original series were also well-known science fiction novelists, including Norman Spinrad, Theodore Sturgeon, and David Gerrold, who got his start writing for the show. Larry Niven wrote an episode for the short-lived animated series in which he injected the Kzinti, an alien race from his *Known Space* stories.

TV science fiction has tended to follow the leader in this regard: the show *Babylon 5*, while different in tone and mood from *Star Trek*, was set on a space station crowded with representatives of dozens of other spacefaring races. This is also typical of science fiction movies as well, the premier example being the *Star Wars* saga. It is also true of the recent *Avatar*, at least implicitly, as humanity finds an example of an intelligent alien race living on an inhabitable moon of a gas giant planet in the nearest star system to our own.

From what I can tell, ideas in the movies and on TV tend to be about 20 years behind science fiction novels and stories. The idea of the inhabited universe seemed to peak in the early 1980s. Again, alien contact is a pervasive theme in science fiction. William Gibson's novel *Mona Lisa Overdrive* marginally involves contact with aliens from Alpha Centauri, although it is completely unimportant to the plot; the novel could have been easily rewritten without it [96].[1] There are many stories in which the existence of alien races is present in the background, so to speak, but is unimportant to the plot: many of Asimov's works fall into this category. The usage seems often to be more a way of signaling the work as science fiction rather than contributing anything of substance to it. However, this is verging on literary criticism, which I promised I wouldn't do.

Among written works, Niven's *Known Space* stories are among the best examples of this idea. Niven has invented many different and fascinating alien races, from the humanoid (the Kzinti and Pak) to the vaguely mammalian (Pierson's Puppeteers) to the nonhumanoid (the sessile but telepathic Grogs and the very alien Outsiders). Again, the idea is quite common: Robert Heinlein's juvenile novel *Have Space Suit—Will Travel* features two alien races at the beginning of the novel (the mother-thing's race and the race of the aliens trying to invade Earth) but leads to the human race being put on trial before an assembly of literally thousands of other intelligent alien races by the end of the novel [114]. I cannot end this discussion without also mentioning Olaf Stapledon's amazing *Star Maker*, a broad "history" of the universe and the quest of its many races to understand its maker [225]. The novel is a classic of science fiction by its best English-language author: descriptions of the alien races are many and varied, including crustacean-like creatures, creatures in the form of intelligent boats, intelligent vegetation, and so on. The science is outdated, but it is still one of the best science fiction works ever written.

The notion of a "crowded universe" stems from the *principle of mediocrity*, otherwise called the Copernican principle: the idea that our planet is a typical planet circling a typical star in a typical galaxy (in perhaps a typical universe....) From everything we know about astronomy, physics and chemistry, this is true, especially now that we know of more than 700 planets circling other stars. However, the "Great Silence" has called this idea into question over the last 20 years or so. Enrico Fermi was perhaps the first person to express the problem: if many races have evolved in our galaxy, then they probably evolved over the course of billions of years. However, given a species capable of traveling among the stars, even at speeds of only a fraction of the speed of light, it would take only millions of years to explore the galaxy. So why aren't they here yet? Or if star travel is impossible, if there are many races out there who want to talk to us via radio communications, why haven't they contacted us yet [132]? SETI has been around in one form or another for fifty years, but the silence has been unbroken.

16.1.2 Alien Life Is Rare but "Out There" for Us to Find

Perhaps there aren't thousands or millions of other races in the galaxy; perhaps there are only one or two other races. The best exemplar of this idea is the novel *The Mote in God's Eye*, by Larry Niven and Jerry Pournelle, which involves the first contact between humans and "moties," humanoid aliens whose runaway population problems cause them to exist in a state of almost perpetual warfare. The reason it takes humanity over a thousand years as a spacefaring civilization to discover the moties has to do with the features of the "hyperspace" drive (invented by Dan Alderson of Caltech to the authors' specifications) [186].

There aren't a lot of novels involving only one alien race out there simply because if you postulate one, you might as well postulate as many as you want.

16.1.3 We Are Alone

One of the earliest major works of science fiction from which aliens were noticeably absent was Asimov's famous *Foundation* trilogy, centering on the decline and fall, then subsequent rebirth, of the "galactic empire." Asimov felt that the absence of aliens was odd enough that in later books in which he combined his robot stories with the Foundation, he went through a lengthy explanation of why the galaxy was devoid of intelligent life apart from humanity.

However, the idea that humanity is the only intelligent race in the galaxy has become, if not the majority position, at least a position among a significant minority, perhaps even a plurality, of science fiction writers since the 1980s. Again, popular culture tends to reflect this. Until the final episode, the reimagined *Battlestar Galactica* featured a universe devoid of alien life. The show focused more on the interaction of the human race with the Cylons, who could be termed an alien species that humanity itself created.

The reason for this gradual change lies in a paradox. Even as scientists have accepted the Copernican principle and have learned that planets, the potential sites of life in the universe, are common, they have also learned that the planets themselves vary enormously from one another, and that the conditions leading to life on a planet are probably very restrictive. Beyond this is the question of the evolution of life on a planet into an intelligent, tool-making civilization capable of space travel, or at least of radio contact with similar civilizations. Most physicists and astronomers tend to feel that once life begins on a planet, the evolution of some species capable of interstellar communication is almost inevitable [239]. The idea is that because evolution proceeds by random changes favoring survival characteristics, and (from the human experience) intelligence and tool using are very much aids to survival, then the evolution of an intelligent species is almost inevitable, given time and the random selection of the fittest.

Evolutionary biologists feel differently. They point out that the evolution of the human race required the convergence of a number of unlikely events. For example, if the dinosaur-killing comet hadn't smashed into the Yucután Peninsula 65 million years ago, mammals might not have become the dominant biological order on the planet.

16.2 MOTIVATION FOR ALIEN CONTACT

If the aliens exist, the second hurdle to overcome is devising a plausible reason for them to contact us. There are good reasons for noncontact: *Star Trek*'s Prime Directive is a good example from science fiction. Historical meetings between technologically advanced civilizations and less advanced ones in Earth's own history underscore the dangers of such contacts. Perhaps the aliens are avoiding contact because they don't want to harm us. Stanislaw Lem illustrated the dangers of such issues [149]. Cosmic timescales being what they are, an advanced civilization may be millions or billions of years older than our own. Any revelations they may choose to send us might be meaningless (like discussing quantum physics with the ancient Phoenicians) or, if practical, extraordinarily harmful (like sending plans for a nuclear bomb to 1939 Europe). However, if the aliens do choose to visit us or talk to us, why might they do so? Some reasons follow.

16.2.1 War with Us

The astronomers tacitly assume that we and the little green monsters would welcome each other and settle down to fascinating conversations. Here again, our own experience on Earth offers useful guidance. We've

already discovered two species that are very intelligent but technically less advanced than we: the common chimpanzee and pygmy chimpanzee. Has our response been to communicate with them? Of course not. Instead, we shoot them, dissect them, cut off their hands for trophies, put them on display in cages, inject them with AIDS virus as a medical experiment and destroy or take over their habitats....If there really are any radio civilizations within listening distance of us, then for heaven's sake let's turn off our own transmitters and try to escape detection, or we're doomed.
—JARED DIAMOND, *THE THIRD CHIMPANZEE* [61, PP. 214–215]

The idea of war with alien races dates back to H. G. Wells's *War of the Worlds*, published in 1898 [248]. The excerpt from Jared Diamond tells you pretty much all you need to know about the conception of the aliens: they are space vampires who survive by exsanguinating human blood and injecting it directly into their veins. They are bent on conquering Earth because Mars, in the novel an older world than Earth, is gradually becoming unfit for life (à la Percival Lowell's ideas). Wells, a passionate socialist, extrapolated the motives and actions of European colonialists in the Americas and Africa for his Martians. The book spawned imitators, especially in movies and TV, where several film adaptations of the novel were made, plus Orson Welles's famous 1938 radio broadcast and movies ranging from *Plan 9 from Outer Space* to *Independence Day*. It's not as popular a theme in science fiction literature as in the movies because of the fundamental difficulties of space travel and the perhaps mistaken idea that advanced civilizations would have moved beyond the need for warfare.

The complementary idea of humans as interplanetary or interstellar aggressors is less common; the recent movie *Avatar* is almost a lone example among science fiction movies. Olaf Stapledon explored this idea in some of his novels published in the 1930s. In particular, *Last and First Men* involves the human race exterminating a race of sentient underwater creatures on Venus. This was because the decay of the Moon's orbit and its ultimate impact on Earth were making Earth unlivable [225]. Despite the scientific inaccuracies, the story provides an interesting psychological study of the Fifth Men, the branch of the human race responsible for the genocide. Stapledon discusses the psychological swings between elation and despair this race faced as it debated the morality of destroying another sentient race to ensure its own survival. His consideration of these moral issues makes Stapledon all but unique among science fiction writers past or present, and his novels are worth reading today. The same idea in a different guise crops up in his novel *Odd John*, about a race of supermen living among ordinary humans. There the debate is over the morality of exterminating the normal human race to ensure the survival of the more evolved species.

This begs the question of whether interstellar warfare is possible. Here the answer is easy: if interstellar travel is possible, then interstellar warfare is possible. The issue has to do with energetics. Let's say that an alien civilization has the capability of creating a starship capable of traveling at 86% the speed of light. The reason for this particular velocity is that the relativistic gamma factor at this speed is 2, meaning that the kinetic energy at this speed is equal to the rest mass energy, Mc^2. Our canonical 10,000 kg starship has a kinetic energy of 9×10^{20} J, the equivalent enegy of 200,000 H-bombs. A few other ways of visualizing this are:

1. The United States uses about 9×10^{19} J of energy in a year, so this is about the total energy use of the United States over a 10-year period.
2. Most asteroids impact the Earth at speeds of about 30 km/s, so if the spaceship impacted Earth, it would have the same effect as an asteroid of mass 200,000,000 kg hitting Earth. This isn't quite in the dinosaur-killer league but it is substantially larger than the comet that impacted Tunguska, Siberia, in 1908.
3. In a prior chapter we saw that turning on the engines for a spaceship of such size would irradiate the Earth with gamma rays equivalent in power to the total energy the Earth receives from the Sun, potentially sterilizing the planet.

The last item is interesting, in that the drive for an interstellar spaceship makes a good weapon all by itself. The first time I encountered this idea was in the story "The Warriors" in Larry Niven's collection *Tales of Known Space*, in which a human spacecraft destroys a Kzinti using its engine, which uses a laser for propulsive power (this is essentially the idea of the photon rocket of chapter 9) [181]. You don't need the drive to be a laser; the propulsion system for any macroscopic spacecraft capable of star travel must be capable of inflicting a lot of damage on a planetary civilization simply by the laws of physics. This make one part of the movie *Avatar* ridiculous. At the end of the movie, the triumphant Na'avi send the humans back to Earth on the spacecraft they arrived in. However, there is no way in which they could have "deweaponized" the starship, which was capable of traveling at two-thirds light speed and was considerably larger than 10,000 kg. The humans could render Pandora uninhabitable simply by turning on its engines. Better for the Na'avi just to kill them all. Jake Sully, of all people, should have spotted this flaw in the plan.

Any star drive is a weapon capable of inflicting horrific, potentially world-sterilizing damage. Any civilization capable of star travel is capable of interstellar warfare. Greg Bear explored this potential in his novels *The Forge of God* and *Anvil of Stars*; in the first, Earth is destroyed by an alien race, and the children of the survivors set out to destroy the aliens responsible in the second [32][34]. The reason the aliens have for destroying the world is a little puzzling; it seems essentially to be a preemptive strike on their part, a pervasive racial paranoia that if they don't destroy us, we will eventually come out and destroy them. I guess this is a possible motivation, the flip side of the worry shown in Diamond's quotation at the opening of this chapter, but it seems over the top. In Niven and Pournelle's *Footfall*, the motivation is that the invading aliens represent a faction that lost a war on their home planet and need Earth as a new home. We see Niven's use of problem limitation here, as any race of beings capable of traveling across the cosmos should be able to pummel twentieth-century Earthlings into submission in a matter of seconds, but the aliens portrayed are not very bright beings created as servants of a highly advanced alien race that killed itself off long before the story begins. The elephant-like aliens are using borrowed technology and have a psychology appropriate to herd animals, allowing humans to dominate them (after a hard fight) despite their initially dropping an asteroid on Earth [187].

Alien invasion doesn't need to be particularly energy expensive; various writers have posited invasion by alien microbes carried across interstellar distances by stellar winds or virtual invasions when alien broadcasts effectively invade our computer systems or give us bad information, as in the movie *Species*. The motive for these attacks is a little hard to judge except for the general rule, "all inhuman aliens are bastards," which seems to be the base assumption of most TV or movies. Humans, of course, broadcast computer viruses essentially for the fun of it, so assuming that alien races would be too evolved to broadcast the interstellar equivalent of ILOVEYOU is probably unwise.

The origin of the idea of the spread of alien life through microbes is probably the "panspermia" theory of Fred Hoyle and Chandra Wickramasinghe, which postulated that life on Earth was seeded by comets carrying the basic building blocks of life. Bacteria and viruses were carried by radiation pressure from one star to another over cosmic timescales, ultimately landing by accident or design (as in the directed panspermia theory of Francis Crick) on a random world, beginning life there if the conditions were right. It's worth taking a little time to evaluate the idea scientifically.

As mentioned in previous chapters, light exerts a force on objects it falls onto. This is the basis of the matter-antimatter (or photon) rocket: the force exerted is given by the formula

$$F_L = kI A/c, \tag{16.1}$$

where I is the total intensity (power per unit area) of the light incident on the object, c is the speed of light, A is the surface area of the object, and k is a dimensionless parameter that describes how effective the object is in scattering light: k depends on the wavelength of the light and the exact shape of the object. If the object is large compared to the wavelength, k will be somewhere between about 1/2 to 2. We'll simply assume it is 1 for now.

If we assume that the microbe being sent through space is spherical, then its area is given by the formula

$$A = 4\pi r^2, \tag{16.2}$$

where r is its radius. Its mass is its volume times its average density (ρ):

$$m = \frac{4\pi}{3}r^3\rho. \tag{16.3}$$

The acceleration of the microbe will be the force exerted on it divided by its mass:

$$a = \frac{F_L}{m} = \frac{3kI}{r\rho c}. \tag{16.4}$$

The smaller the object, the greater the acceleration. In Earth's orbit the solar constant (intensity of light from the Sun) is 1,360 W/m^2; if we assume that our microbes have a density the same as water (1,000 kg/m^3) and a radius of 1 μm (=10^{-6} m), we get an acceleration of about 0.01 m/s^2. This doesn't sound so high, and of course, the acceleration drops as it gets farther from the Sun, but with this acceleration over a distance of 1 AU, the speed of the bacterium will be 64 km/s—greater than the escape speed from the Solar System. It isn't going anywhere fast, but it could make it to the Alpha Centauri system in about 20,000 years. One plausible invasion scenario would be genetically engineered bacteria or viruses, or even the basic genetic material needed to seed other planets with higher-order life, sent out by an alien race to colonize other worlds. This could be done at a tiny fraction of the cost of sending starships containing fully sized aliens: a bacterium moving at 86% of the speed of light has a kinetic energy of only about 400 J. Now, there isn't any way I know to slow it down once it reaches its destination (though there might be). This might be the means by which the unseen alien intelligences managed the Chtorr invasion of the Earth in Gerrold's series, although it hasn't been revealed yet. Fast or slow, this invasion strategy is much more plausible than almost any other scenario because of the small amount of energy needed.

16.2.2 Trade or "Enlightenment"

> When you trade among the stars, *there is no repeat business.*
> —LARRY NIVEN, "THE FOURTH PROFESSION," IN *ALL THE MYRIAD WAYS*

The issue of commerce between the stars is pretty simple: energetics and time effectively prevent it, or at least prevent physical trade. Consider the following: even traveling at 86% of the speed of light, travel to Alpha Centauri will have a round-trip time of 10 years. And that's the nearest system. The cost is also horrific: the kinetic energy of anything with a mass of 1 kg moving at that speed is 9×10^{16} J. At current energy costs of about 10 cents per kW-hr this corresponds to about \$300 million per kilogram. This isn't the entire cost, however, as the relativistic rocket equation tells us that we need 1.4 kg of antimatter (combined with an equal amount of matter) to give 1 kg of payload this velocity. Plus more to slow down to a stop, reverse course with the goods, and slow to a stop again on return. It is hard to imagine *anything* that could be traded at a profit given the fundamental problems. This is where the fundamental premise of the movie *Avatar* breaks down: the cost of unobtainium, available only on Pandora, will probably be at least an order of magnitude higher than the quoted value of \$20 million per kilogram.

The exchange of information is orders of magnitude cheaper. Communication between alien civilizations via radio waves is faster and cheaper than any exchange of material goods: radio travels at the speed of light, faster than any material object can travel. Cocconi and Morrison showed in the 1950s that it would be possible to detect Earth's radio transmissions, using then current radio telescope technology, from a distance of more than 10 light-years [60]. So communications between alien civilizations is possible, but is there anything we would want to say to them? I discussed this above: it is possible that aliens wouldn't want to talk simply because they wouldn't have much to say to us. However, merely saying "hi" and nothing more would immediately become the most important radio message ever received.

16.2.3 Mars Needs Women!

> The successful cross between a human and a Vulcan ... is about as likely as the successful mating of a Vulcan and a petunia.
> —CARL SAGAN, 1968 *NEW YORK TIMES* ARTICLE

This is more of a joke than anything else: any competent science fiction writer understands the issue that interspecies "relations" are unlikely. The genetic information for most species on Earth is stored in deoxyribonucleic acid (DNA); it is unlikely that this is the only possible genetic molecule, so there is almost no chance that life evolved on another world would be "compatible" with ours. Even if it was DNA based, aliens would represent different species from humans. As Larry Niven put it in his essay "Man of Steel, Woman of Kleenex," sexual relations between a human and an alien is what Tom Lehrer once called "animal husbandry" [178].

There is also no reason for them to look much like us. TV shows feature humanoid aliens because it is easier to make human actors look like humanoid aliens than protoplasmic blobs. Perhaps the Anna Karenina principle comes into play here, too: there may be issues that force intelligent, tool-wielding aliens to look more or less humanoid. Until we find examples, of course, we can't know.

16.2.4 Utterly Alien Motives

A large genre of science fiction concerns alien contact in which the motives for contact are unknown and to a large extent unguessable. The authors of these stories tend to be non-U.S. or non-UK writers, many of them from the former Soviet bloc countries. In the Russian novel *Roadside Picnic*, by Arkady and Boris Strugatsky, the alien contact is invisible in addition to being inscrutable [230]. In this novel, the "Visitation" has altered several areas of Earth: in them, the laws of physics break down, miracles happen (usually deadly ones), and there are treasures for intrepid explorers to find. The aliens are entirely offstage (if they were ever there in the first place). The only evidence that they ever came is the "Pillman radiant," an imaginary line in space along which the Visitation seems to have come. Dr. Pillman, the discoverer of the radiant, likens the Visitation to a roadside picnic in which visitors to a picnic site leave their trash without bothering to pick it up, but it isn't clear that this is the motivation for the Visitation. Stanislaw Lem in *Microworlds* gives a more honorable motive for the Visitation, but it isn't clear that his explanation is correct, either [150]. I suspect that *Roadside Picnic* was a primary motivation for the excellent TV miniseries *The Lost Room*, but cannot prove it; the "Event" of that series is very similar to the Visitation of the novel.

Stanislaw Lem has two novels premised on alien contact with unknowable motives, *Solaris* and *His Master's Voice*, and a third, *Fiasco*, in which the tables are turned: humanity attempts to contact an alien race but our motives are completely misunderstood [151]. In *Solaris* an intelligent planet attempts experiments on astronauts investigating it, for completely unknown reasons. [148]. In *His Master's Voice*, scientists intercept an alien message and are completely unable to decipher it; various attempts lead to interesting discoveries, but the content of the message remains completely unknown [147].

Philip K. Dick is one of relatively few American writers who have attempted stories constructed on unknowable alien motives. In *Galactic Pot-Healer* the protagonist (a "pot-healer," or ceramics mender) is dragged into a conflict on a distant planet where he understands only the vaguest of the motives or powers of the beings fighting one another to raise a sunken cathedral [68], I find stories like this plausible, in that it is very difficult to communicate with members of species as close to humans as chimpanzees. What luck would we have with aliens? But putting all this aside, let's get down to Earth and think about how likely it is for us to be able to make contact, whether we can understand them or not.

16.3 DRAKE-EQUATION MODELS AND THE MATHEMATICS OF ALIEN CONTACT

If we were to receive a greeting from an alien civilization tomorrow, we could be pretty confident it was within 50 light-years of us. Our world has been broadcasting radio signals for only about a century, so to receive a signal from aliens would indicate they found our signal and sent us back a reply within that time. This implies a certain density of alien civilizations in our galaxy of roughly one per every $(50 \, \text{LY})^3$, or $8 \times 10^{-6} \, \text{LY}^{-3}$. The Milky Way Galaxy is a spiral galaxy composed of 3×10^{11} stars, some 1×10^5 light-years across and 1,000 light-years thick, so the total volume is approximately that of a cylinder with radius $R = 5 \times 10^4$ light-years and thickness 1,000 light-years; this implies that the total volume is $\approx 8 \times 10^{12} \, \text{LY}^3$. If such civilizations are distributed uniformly throughout the galaxy (which is not necessarily a good assumption), then the reception of such a message would imply about 64 million advanced civilizations currently in the galaxy. This seems like an incredibly large number. Of course, the fewer civilizations there are, the longer it will take to talk to them.

Let's make a simple model for this. Let's hypothesize that the average number of alien civilizations currently in our galaxy can be expressed as

$$N = gL. \tag{16.5}$$

Here, g is a "generation rate," which indicates the number of galactic civilizations that achieve a technological level suitable for interstellar communication; this is a certain number per year being "born." N is the number of such civilizations in the galaxy at any given time, and L is the lifetime of the civilization. Let me give an analogy for what is going on here: Imagine standing in a dark meadow in the middle of the countryside in summer. All around you, you see fireflies flicker on and off. How often do you see more than one firefly lit up? That depends: if a lot of fireflies on average "light up" every second, or if they stay lit for long periods of time, then your odds are good that you will see more than one. If, on the other hand, very few light up every second and they stay lit up only briefly, then the odds aren't good. The number of fireflies lighting up every second corresponds to g; the amount of time they stay lit up corresponds to L. Intelligent life in the cosmos may be akin to those fireflies—brief flashes lighting the darkness.

All three quantities, N, g, and L, are unknown. The astute reader will recognize this as a version of the Drake equation. In 1960 Frank Drake proposed a statistical model to estimate the number of alien civilizations in the galaxy. In his original version of the equation,

$$g = G f_p n_e f_l f_i f_c. \tag{16.6}$$

It's worthwhile considering the variables here:

- G: the rate at which stars form. This is the only variable that is known with any precision: $G \approx 7$ per year.
- f_p: the fraction of all stars in the galaxy with planets. In 1960 there were no known exoplanets; now there are more than 700 known ones. The value

of f_p is probably somewhere between 0.05 and 0.2 (5% and 20%) based on the statistics of stars with planets. As discussed in previous chapters, the probability of a given star having detectable planets increases with increasing metallicity, and is also a function of stellar class.

- n_e: the average number of Earth-like planets circling the star. This figure is highly problematic. As discussed earlier, a planet being "Earth-like" depends on a number of factors, some of which are interrelated. Depending on how you define it, there are between one and three Earth-like planets in our own Solar System, but the average number per solar system is likely to be much lower.
- f_l: the fraction of Earth-like planets that develop life. This and the next two variables are completely unknown. It might be high, as Earth developed life only about 700 million years after it formed.
- f_i: the fraction of planets developing intelligent life.
- f_c: the fraction of intelligent life that develops technology able to communicate over interstellar distances (i.e., radio).

Needless to say, the equation and its interpretation have engendered much controversy. In response to criticism, Drake has called it a way to "organize our ignorance." The formula is valid for estimating the number of intelligent alien races only if the variables are statistically independent of one another. This is unlikely to be true. In particular, when we look at n_e, what exactly do we mean by an Earth-like planet? That number, n_e, could be estimated by a mini-Drake equation of its own: we could estimate the average number of planets in the habitable zone times the number whose orbital eccentricities are small enough times the number that are the right size times. . . .

In any event, we don't know most of the factors that go into g. We also don't know L. Remember, L is the lifetime of the "typical" advanced civilization. However, "advanced" means "being able to receive and transmit communications over interstellar distances"—otherwise we couldn't communicate with them. On Earth we have had this capability for less than 110 years, depending on how you define it. We've been listening for alien signals for only about 50 years. This probably represents a lower bound on L, but it isn't clear what an upper bound should be.

Michael Shermer in an article in *Scientific American* estimated the average life span of a human civilization as 420 years; these civilizations were all preindustrial, but if this is a reasonable estimate for the life span of a technological civilization, there is no hope of contacting aliens [218]. There has been heated debate over this article on the web, but Shermer's reasoning isn't stupid. For one thing, any advanced technological society implies a wealth of energy. The generation of energy in our society rests almost entirely on nonrenewable resources, which will probably be depleted in less than a century if their use continues at the same rate as today. We'll take up the issue of how long human civilization can last in later chapters. For now, is there anything at all we can say about the possibility of alien contact, given our lack of knowledge of all of the variables above?

Well, first of all, we want the product $gL > 2$, so that there is someone else to talk to. This is not a hard-and-fast rule, as these numbers represent averages, but it seems like a reasonable criterion. Unfortunately, this is a necessary but not sufficient condition. We also need to be able to communicate, which will be impossible to do within the lifetime of our civilization if the alien worlds are spread too far apart.

If galactic civilizations are uniformly distributed, then we can imagine them as being spread over the volume of a cylinder with a very high aspect ratio (i.e., 100:1) between radius and height. Therefore, if these civilizations are separated by distances larger than about 1,000 light-years, it makes more sense to discuss them in terms of the number per unit area spread over a disk representing the galaxy with radius $R = 50{,}000$ light-years. If we send a radio signal from Earth and expect it to be intercepted by aliens after a time t, then we can relate N to t as follows:

$$N \frac{(ct)^2}{R^2} \geq 1. \tag{16.7}$$

This is a straightforward probability argument: the expected number of civilizations that receive the signal is equal to the area the signal spreads over (ignoring the thickness of the galactic disk) multiplied by the area density of such civilizations. (From here on, I'll work in units in which $c = 1$, to make life easier).

If we want to receive a signal in return, then $t \leq L/2$. That is, if our advanced technological civilization has a finite life, the aliens had better

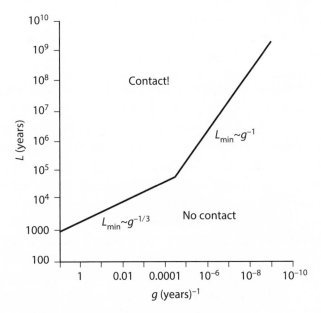

Figure 16.1. Minimum values of L if we want to make contact.

be close enough to signal us back before it ends. From this, we can put a lower bound on the lifetime required for us to be able to talk to the aliens. Using $N = gL$, we can derive

$$L > \frac{(4R^2)^{2/3}}{g^{1/3}} \approx 2000 \, \text{yr} \, g^{-1/3}. \tag{16.8}$$

So we have two criteria:

$$L > \frac{2}{g}$$

and

$$L > \frac{2{,}000}{g^{1/3}}.$$

Both criteria must be satisfied in order to be able to talk with the aliens. Figure 16.1 shows what this implies. Note that g decreases to the right.

The graph is separated into two regions. Above the line, contact is possible. There are enough aliens out there to talk to and our civilizations will endure for a long enough time to make contact. Below the line, contact is unlikely, as either there are no other civilizations in the galaxy or they are separated by such a large distance that contact will not be made within the lifetime of the civilization.

For low values of g, the harder criterion to satisfy is the first one; the odds of having any other civilizations in the galaxy at a given time are low no matter what you do. However, as g gets larger, for a value of g around 3×10^{-5} per year, it is now harder to satisfy the criterion that the civilization must last long enough for a conversation to happen. It is very likely that $g < 10$ per year, as the stellar generation rate is less than this, and all the other factors going into g tend to decrease it. The implication is that no matter what, the minimum lifetime for advanced civilizations is about 1,000 years if we want contact to be even a bare possibility. This leads us into the final section of the book: how long can an intelligent species hope to last?

Note

1. Gibson is an interesting example of an author who has gradually removed many of what people think of as science fiction trappings (alien contact and space travel, to name two things) from his novels while still keeping a very science fiction feel to his writing.

PART IV

YEAR GOOGOL

CHAPTER SEVENTEEN

THE SHORT-TERM SURVIVAL OF HUMANITY

17.1 THIS IS THE WAY THE WORLD WILL END

This section poses a simple question: how long can humanity last? Here I am going to throw caution to the winds: This is not a question that I will try to cast in terms of realistic bounds; instead, I will simply speculate wildly as to how long the laws of physics will let us last. In doing so, I examine the different ways in which the human race can become extinct. I move forward in jumps: first considering all the many, many ways in which we can do ourselves in during the next century or so, then taking up more cosmic catastrophes. From the laws we understand today, if we get lucky, the species could last as long as a googol years.

17.2 THE SHORT-TERM: MAN-MADE CATASTROPHES

So, whither humanity? I am greatly indebted to the books *Last and First Men* and *Star Maker* and their author, Olaf Stapledon, a British science fiction writer of the 1930s. Both are tour de force looks at the long-term history of life in the cosmos. *Last and First Men* examines the future of the human race out to a time trillions of years in the future when our descendants have evolved into hyperintelligent giants with telescopes on their heads living on the planet Neptune (it makes sense in context). *Star Maker* is even more audacious: it attempts to follow the evolution of all life everywhere in the universe, to a time thousands of eons past the end of the human race, and life's (ultimately fruitless) attempts to make sense of the universe and its creator. Along the way, Stapledon explores the myriad ways in which humanity and other species come to catastrophe and recover (or not) in their struggle upward. The novels, despite being written 80 years ago, are the best examples of "thinking big" in all of science fiction. Among other things, Stapledon invented the idea later christened the Dyson sphere.

Of course, to survive past the end of all the stars, we need to survive past the end of the century. In the short term, mankind is most likely doomed by... Mankind.

17.2.1 Nuclear War

> But we can be thankful and tranquil and proud,
> That Man's been endowed with a mushroom-shaped cloud,
> And we know for certain that one lucky day,
> Someone will set the spark off,
> And we will all be blown away!
> —THE KINGSTON TRIO, "MERRY LITTLE MINUET"

An all-out nuclear was the odds-on favorite for the demise of the human race when I was a boy. Countless science fiction novels and stories deal with all-out nuclear war, usually between the United States and the late Soviet Union. Of these, *On the Beach* by Neville Schute is one of the best. It deals with the aftermath of a nuclear holocaust, with a number of survivors waiting for the end as radioactive fallout gradually kills off all life. Most books of this kind were written roughly between 1950 and 1970, when Cold War tensions were at their peak. There are also a number of science fiction novels from this time about aliens coming to Earth and saving us from a nuclear holocaust. The best of these is *Childhood's End*, by Arthur C. Clarke, because it presents a nifty twist on the story: the aliens save us, but they are saving us for something else.

I won't say anything more because if you haven't read this novel, you need to. It's a bit dated, but it still packs a kick like a mule.

You used to see bumper stickers that read, "One nuclear bomb can really ruin your day." The only military use of atomic weapons was at the end of World War II, when the United States used two 10 kt atomic bombs to destroy the cities of Hiroshima and Nagasaki. Estimates vary, but the bombs were responsible for the death of up to 100,000 people in each city. The center of each city was completely destroyed. The human misery caused by the bombing and the radioactive fallout were also staggering; there isn't enough space here to do it justice, but read the end of the book *The Making of the Atomic Bomb* by Richard Rhodes for details if you want [197].

At the height of the Cold War, the combined strength of the nuclear arsenal of the United States and Soviet Union was roughly one million times the power of the Hiroshima bomb. There is an interesting thing about this, however: the radius of complete annihilation of a bomb doesn't scale as the bomb energy but as its square root. For example, the 10 kt bomb that destroyed Hiroshima could probably wipe out Manhattan, which is about 1 mile long. A 1 megaton hydrogen bomb, 100 times more powerful, wouldn't wipe out an area with a radius of 100 miles but one with a radius of only about 10 miles, about the size of the five boroughs. Therefore, the combined nuclear arsenal of both powers could directly destroy an area roughly 1,000 miles in radius—big, but not Earth-shattering. I don't want to trivialize the suffering that would be caused by such a holocaust—far from it. As a teenager living in the D.C. suburbs, I used to stay up nights worrying about the imminent threat of nuclear holocaust. The bombs would be targeted at the major cities where the bulk of humanity lives, so the destructive power would be devastating. Cold War strategists used to estimate the numbers of people killed in terms of "megadeaths"; Dr. Strangelove of the eponymous film is a parody of one of these cold-blooded types. Death estimates for an all-out nuclear war would start in the region of ten million, going to an upper range of over 100 million people killed immediately. And that is to say nothing of the lingering death by radiation poisoning of an equal number of people.

Yet even given all of this horror, a nuclear war wouldn't directly kill off everyone on the planet. There are six billion people alive today;

perhaps 1 to 10% would be killed off directly or from fallout. The biggest danger of a nuclear holocaust is probably not the destructive power of the bombs themselves, great though it is, nor that of the nuclear fallout, horrifying though that is. It is the danger of nuclear winter.

Consider the following: the destructive energy in a 10 MT H-bomb is about 4×10^{16} J. In addition to the heat and shock wave produced by the blast, a large amount of dust and other particulate matter can be lifted high into the stratosphere [97]. The stratosphere, the highest region of the Earth's atmosphere, begins roughly 100 km above the surface of the Earth. The energy required to lift 1 kg of material into this region is given by the formula

$$E = mgh = 1\,\text{kg} \times 10\,\text{m/s}^2 \times 10^5\,\text{m} \approx 10^6\,\text{J}.$$

If only 1% of the bomb energy went into lifting particulates into the stratosphere, then we could expect about 4×10^7 kg per megaton lifted there. This estimate is pretty crude, so we can't expect better than an order of magnitude agreement. Most papers on global warming assume a value of about 3×10^8 kg per megaton. In addition, soot from burning cities and forest fires would add to the total [200]. Such particulates take a long time to settle out of the atmosphere, possibly up to several years, and they act to block sunlight from reaching the Earth. Even a small nuclear war, such as a limited exchange between India and Pakistan, could lower global mean temperatures by nearly 1 °C, resulting in large-scale changes in agriculture and possible famine.

This scenario, refered to as "nuclear winter," will come back in an altered form in chapter 21, when we talk about the prospects for the long-term survival of humanity. However, if the global climate were altered significantly by a massive nuclear exchange, there is a good chance that a very large fraction of humanity would die as a result of the disruption of agriculture.

17.2.2 Global Warming

At the present time, however, the world is in danger of getting too hot, not too cool. Since the breakup of the Soviet Union, fears of all-out nuclear war have more or less died away; there is always the danger of a rogue state or a terrorist group getting its hands on a nuclear bomb, but there is less of a chance that the world will end in a huge holocaust. The world's nuclear arsenal has been reduced by a factor of three since the breakup of the Soviet Union. In 2010 the Doomsday Clock in the *Bulletin of the Atomic Scientists* was pushed back from five minutes before midnight—the witching hour—to six. Then in 2012 it went back to five minutes before midnight because of the threat of global climate change.

Most people have some idea what the phrase "global warming" means. There are in fact three related terms to be understand: "global warming," "global climate change," and the "anthropogenic greenhouse effect." In a nutshell, the Earth's mean temperature seems to be rising as a result of the large-scale production of carbon dioxide (CO_2) and other industrial gases by our civilization; this has been going on for about the past 200 years. The International Panel on Climate Change (IPCC) in its 2007 report estimated that the world's mean temperature will rise somewhere between 2 °C and 6 °C by the year 2100, with unpredictable consequences for human civilization [223][1].

A few words before I go on to discuss the science behind anthropogenic (man-made) global climate change. Some seem to think that this theory is somehow controversial, and that not everyone is convinced that it is true. Hogwash! The people who matter, which is to say the people who spend their lives studying this theory (i.e., climatologists), are of almost uniform opinion on the subject. Global warming is real, and it is dangerous. At this point the processes of global warming cannot be reversed (though the effects can be partially mitigated). It is inevitable that the world's temperature will rise significantly over the next century. The disagreements are only about the details: how much, which parts of the world will be most affected, and so on. The theory is controversial in the same way in which the theory of evolution by natural selection is controversial: accepted by scientists, but not accepted by people who are either scared of the implications of the theory or have a stake in the theory not being true.

Of course, global climate change has been incorporated into science fiction novels. Kim Stanley Robinson has written a trilogy dealing with the effects of global climate change. The first novel in the trilogy, *40 Signs of Rain*, looks at the effects of the Earth passing a "tipping point," after which civilization begins to deal with the effects of rapid climate change [199].

We have already discussed the mathematics behind the greenhouse effect on planetary temperatures in chapter 14: suffice it to say here that certain gases in Earth's atmosphere act to trap heat near the surface of the Earth, raising Earth's mean temperature though the blanketing effect of these gases in exactly the same way a thick blanket or jacket keeps you warm. The natural greenhouse effect in general is good. Earth would be some $30°$ C colder than it is now without it, cold enough that ice would cover most of the planet most of the year. However, since James Watt invented the first commercially viable steam engine in 1776, humanity has burned coal and later oil to provide energy for its rapidly expanding industrial civilization. This has increased the CO_2 concentration of the atmosphere by about 33% over its preindustrial level. The first scientist to understand that increasing human-created greenhouse gases would lead to increasing mean global temperature was the 1903 Nobel Laureate in Chemistry, Svante Arrhenius [141]. His initial work marks the first important theoretical study of climatology; he did what modern climatologists do, modeling the effects of increased greenhouse "forcing" using a sophisticated geographical model. Unlike modern climatologists, he didn't have access to modern computers. Calculating the effects of greenhouse forcing by hand took him several years. However, his estimate of warming of a few degrees Celsius is similar to modern estimates. He felt that this would be a good thing in making the world's climate more tropical; perhaps his optimism stemmed from the fact that he lived in Sweden [141].

The IPCC predictions are based on sophisticated computer models that do the same sort of calculations Arrhenius did at the turn of the last century. The models divide the world into a three-dimensional grid; horizontal gridding is in squares of a few kilometers on a side, while the atmosphere is considered in terms of vertical layers of about 1 km height. The models look at heat input from the sun, the amount reflected from the surface of the Earth and from atmospheric aerosols and clouds, and the amount trapped by greenhouse gases, plus transport across different layers due to air motion [157, p. 68]. The models are extensively tested against each other and against predictions made to local climate conditions; one test used by most modelers is to examine cooling of the Earth as a result of volcanic eruptions, such as the 1993 Pinatubo explosion and the 2010 Eyjafjallajökull eruption in Iceland.

In chapter 14 we presented a simple "two-layer" model to estimate the mean temperature of a terrestrial planet, including the greenhouse effect. The layers considered are the planetary surface treated as a whole and the atmosphere, again treated as a whole. There are four inputs to the formula: stellar luminosity, expressed in terms of the luminosity of the Sun; planetary distance, expressed as a multiple of the distance of Earth from the Sun; mean planetary albedo A, the fraction of light reflected from the planetary surface ($A = 0.3$ for Earth), and the fraction of reradiated heat trapped by the atmosphere, f. For Earth, putting the numbers in the formula results in a mean planetary temperature of

$$T = 278 \, \text{K} \times \sqrt[4]{\frac{1 - A}{1 - f}} = \frac{254 \, \text{K}}{\sqrt[4]{1 - f/2}}. \tag{17.1}$$

This highlights the importance of the greenhouse effect for life on Earth. Without warming from atmospheric greenhouse gases, the case when $f = 0$, the Earth's mean temperature would be a chilly 254 K, or about $-19\,°C$. We can calculate f from Earth's mean temperature, 288 K, indicating $f = 0.77$ in this crude model. Of course, this doesn't tell us what f is due to; *that* is a pretty involved calculation, as one needs to know the infrared absorption bands of all the important greenhouse gases, which also change with atmospheric concentration and pressure. Modeling these absorption bands is one factor in how complicated the climate modeling programs are. The parameter f depends on greenhouse gas concentrations but is not an additive quantity in this if concentrations become too high. Also, of course, different gases are more or less effective at trapping radiation; for example, methane is twenty times more effective by weight in trapping radiation than CO_2 is. However, there is about 300 times more CO_2 in the atmosphere than methane, so CO_2 contributes about ten times more to global warming than methane does. Water vapor is the most important greenhouse gas, representing some 80% of the total effect, with CO_2 being the second most important contributor at roughly 10%. Other types of greenhouse gases contribute a few percent or less.

I present a simple model to estimate the change in temperature due to a change in the fraction of heat trapped by the atmosphere. Does a 1% increase in absorption mean a 1% temperature increase? No We can use calculus and equation (17.1) to show that

$$\frac{\Delta T}{T} = \frac{1}{4}\frac{\Delta f}{f}. \tag{17.2}$$

If f increases by 4%, the global mean temperature will increase by about 1%, or roughly 3 °C, since the mean temperature is around 300 K. A very crude estimate of the anthropogenic greenhouse effect can be made as follows: greenhouse warming by CO_2 represents about 10% of the total. Since CO_2 atmospheric concentration has increased by about 25% since 1800 CE, total greenhouse forcing has increased by about 2.5%. The equation thus predicts an increase in world temperature of about 0.6%, or roughly 2 ° C. This isn't bad for a very crude estimate; mean world temperature has risen by somewhere between 0.75°C and 1.5°C since 1800, or about a factor of two off from our crude estimate.

The effects of global warming are unforseeable in detail but aren't likely to be good for humanity. Predictions include more and more severe droughts, particularly in areas such as California that are already suffering. Also, we will see more frequent and more violent hurricanes, with corresponding billions of dollars in property damage. The melting of the polar ice caps, sea-level rise, and other unsavory occurrences can also be expected. Major drought in California alone would be very bad news. The worst-case scenario would make large portions of the state uninhabitable and unsuitable for agriculture. Since fully 30% of all of America's agriculture is in California, this could be a great disaster, with escalating food prices being the least of the problems we could face.

Lest this seem unthinkable, the United States has experienced a similar if smaller-scale ecological disaster in the last century: in the Oklahoma dust bowl disaster of the 1930s. Overuse of farmland in the western states and a series of droughts led to large-scale topsoil erosion, crop failure, and mass migration of farmworkers out of the worst-affected areas. Combined with the Great Depression, framland drought led to vast human suffering. If I may refer to a non-science fiction novel, John Steinbeck's *The Grapes of Wrath* is a fictionalized but realistic account of one family dealing with these problems [226]. Recently, with a better understanding of climatic variations, archaeologists have found that climate has played a significant role in history. Jared Diamond's *Collapse* and Elizabeth Kolbert's *Field Notes from a Catastrophe* have sections reflecting on how climatic change led to the collapse of various historic and prehistoric societies [66][141]. If it's any consolation, we may be running out of the fossil fuels whose consumption leads to such atmospheric warming.

17.2.3 Hubbert's Peak

> For the petroleum industry, the last century has been a period of bold discovery and adventure. Whole petroleum provinces analogous to the continents have been discovered and partly explored; a few tens of very large fields, corresponding to the large islands, and hundreds of small fields, the small islands, have been discovered. But how far along have we come on our way to complete exploration?
>
> —M. KING HUBBERT, *NUCLEAR ENERGY AND THE FOSSIL FUELS*

When I was a lad of eight, I remember waiting in the car for an hour with my mother to buy gas. This was back in 1973, during the short-lived OPEC oil embargo. Many of the Persian Gulf countries had stopped selling oil to the United States because of U.S. support of Israel during the Yom Kippur war. The government instituted gas rationing based on the last digit of the license plate of the car one drove. The embargo didn't last long, as the United States and the Persian Gulf countries are in a "co-dependent" relationship. But at one point, the United States was the largest oil-producing country in the world, and one of the biggest oil exporters. Now we are seventeenth on the list of oil-producing countries and the single largest oil importer. We produce only about one-fifth of the oil we consume. What happened?

We ran out. Well, not exactly; we're not out of oil yet, but we have run low. It's a simple idea: you can't pour 11 gallons out of a 10-gallon jug. The year 1972 was the first year in which the United States produced less oil than it consumed, making it vulnerable to the OPEC oil embargo.

Everyone, including the OPEC countries, was pretty surprised at how well it worked, or at least how much confusion and panic it caused. And no one saw it coming.

Well, not quite no one. M. King Hubbert, a geophysicist working for Shell, saw the oil deficit coming a long time before it happened. In 1956 he published a visionary paper, "Nuclear Energy and the Fossil Fuels," in which he compared the search for oil during the late 1800s and early 1900s to the age of exploration in the late Middle Ages, when Western civilization put together a map of the world. His point was that during the early age of petroleum exploration, just as in the earlier age of exploration, one couldn't be sure what one would find. No one knew where the oil was or how much there was of it. But as of 1956, Hubbert felt that enough of the Earth's great oil deposits (particularly in the United States) had been located to allow a reasonable estimate of the world's oil supply. He wrote, "What we seem to have achieved is an abundance of detailed charts of local areas with only an occasional attempt to construct a map of the whole world which, despite its inherent imperfections, is still neccessary if we are to have an approximate idea of where we are now and where we are going" [126].

The two key ideas Hubbert drew on in his paper were, (1) there is a finite amount of the fossil fuels (coal, oil, and natural gas) and (2) world civilization uses these resources at an ever-increasing rate. Indeed, their usage is at an exponentially increasing rate. Historically, fossil fuels were very little used before about 1800—that is, at the beginning of the Industrial Revolution. However, as the world became more industrialized, this use rate increased, at one point in the 1950s being something like an increase of 9% per year. This worked as long as we had enough oil in the ground to sustain such increases. However, as Hubbert pointed out, the total amount of oil in the ground could be estimated with relatively good accuracy: the United States had some 200 billion barrels, while the total world supply was roughly two trillion barrels [127].[2]

Current world use rate is roughly 80 million barrels per day, or about 30 billion barrels per year. Of this, the United States consumes about one-fourth of the total. Currently, the rate at which oil is produced in the world matches the demand. However, an oil industry rule of thumb states that the maximum amount of oil that can be extracted from a well per year is roughly 10% of the remaining oil left in it.

Therefore, as oil wells become depleted, not only is there less left, we can take out less every year. So we are faced with exponential growth in oil use until the day comes when the demand for it becomes greater than the supply, at which point the rate at which we use oil must begin to decline. Again, the idea is simple: if a resource is finite, at some point you have to start using less of it. Exponential growth rates are unsustainable.

One way to describe this is via a logistic function. This is the function satisfied by the differential equation

$$\frac{dx}{dt} = \gamma x(1 - x), \tag{17.3}$$

where γ represents the (yearly) growth rate characterizing the curve. Here, $x(t)$ represents the total fraction of the resource that has been used up by time t; it goes from 0 to 1, 1 meaning that all of it has been used up. We are actually more interested in the rate dx/dt, as this is more relevant to the economics of oil production. Without even solving the equation, we can see that the usage rate will reach a maximum when $x = 1/2$, with a value of $\gamma/4$.

This is the simplest possible model describing such a situation because the factor γ is tied to demand for oil or whatever natural resource is being consumed. Thus, γ was essentially zero before about the year 1900, when the world switched from coal to oil as its primary fuel source. Because of this, it will generally be a function of time. In particular, as the oil runs out, if the world switches to another type of energy source, the demand rate will decrease significantly again, maybe to zero, more likely to a lower but non-zero value. For now I'll assume that γ is a constant value and see what results we get.

Current world oil use is roughly 30 billion barrels per year and the world used roughly one trillion barrels between the year 1900 and 2010. Therefore, at the present time x is approximately .43 (i.e., we've gone through slightly less than half the oil in the ground), and we can work out that $\gamma \approx 0.12$ from this data. We can back-project from this figure to see whether this figure represents a realistic model for world oil usage. Please note that these values are very approximate, for reasons I discuss below.

The solution to the logistical equation is

$$x(t) = \frac{e^{\gamma t}}{1 + e^{\gamma t}} \tag{17.4}$$

and the fractional use rate from this model is

$$\frac{dx}{dt} = \gamma \frac{e^{\gamma t}}{(1 + e^{\gamma t})^2}. \tag{17.5}$$

Note that on this scale we have a maximum value of the use rate when $t = 0$. Also, in principle, the assumption is that the world has been using oil indefinitely into the past, which isn't true. Industrial society only began using it in large quantities around the year 1900 CE. However, in practice the model has such a small use rate for times more than 100 years ago that it doesn't matter. We can then fit this to the total world use rate by multiplying by the total amount of oil in the world, Q:

$$R(t) = Q\frac{dx}{dt} = Q\gamma \frac{e^{\gamma t}}{(1 + e^{\gamma t})^2}.$$

I assume a value $Q \approx 2.3$ trillion barrels, or 2,300 billion barrels.

The predictions of the model are interesting. Figure 17.1 shows the results of this calculation. Using these numbers, and looking at our current world usage rate of 30 billion barrels per year, we are approximately sixteen years away from the peak, at which point the world will be using about 70 billion barrels per year. Thereafter, there is a rapid decline. Sixteen years after the peak, world usage is back down to 30 billion barrels per year and falling exponentially. However, this curve doesn't fit the actual data from the past very well, which is not too surprising. The exponential rise part of the curve indicates a 12% growth rate per year, which is almost impossible for any society. Two other curves are shown: one made using $\gamma = 0.09$ and a second using $\gamma = 0.06$. Interestingly, all curves pass the 30 billion barrel per year mark at around 10–15 years before the peak, but the peak production rate is proportional to γ, and therefore is only about 35 billion barrels per year for the lowest value. To compensate, however, the oil lasts longer.

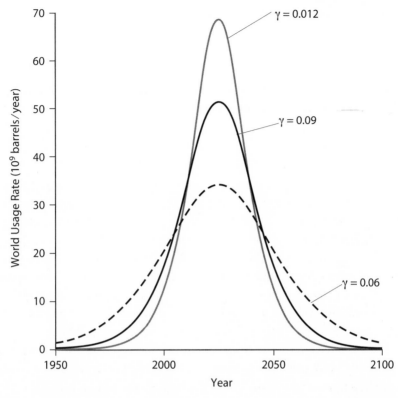

Figure 17.1. Logistic graph showing a mathematical model of world oil use. The graphs had been scaled so that oil usage is approximately 30 billion barrels per year in 2010.

One point about all models: they are extremely dependent on the assumptions one makes going in. This is one major problem with economic models, which drives physicists like me to despair. Most economic models are based on shaky assumptions regarding supply and demand. Most of the parameters are fit by hand, as in the example given above rather, than derived from first principles. Also, there is nothing particularly special about a logistic curve. Many different mathematical models predict similar things. Hubbert, to his credit, understood this. If you go back to his original 1956 paper and his 1982 survey article in *The American Journal of Physics*, you are struck by how bright he was. His assumptions are based on rather general mathematical principles. Many different curves look globally similar to a logistic curve, the difference lying in the fine details. Other examples of such sigmoidal curves are the error function and the Gompertz curve.

So oil will, perhaps, begin running out in about ten years, maybe sooner. A recent survey by the Kuwaiti government indicated that the peak might occur as soon as 2014; other estimates place it nearer 2020. What happens after that? According to whom you believe, either not much or Doomsday.

Lets look at Doomsday scenarios first. Imagine that I could wave a wand and utter the invocation "Petrolevaporatus," magically destroying all the world's oil supply, drying up all the gas in every gas tank in every car, emptying every oil well on the planet. If that happened, about 99% of the U.S. population would die. Period. Think about it: you don't grow your own food—at least, not enough to feed yourself. Only 2% of Americans nowadays are farmers, and the bulk of our food is grown on large industrial farms. Our major food crop is corn, mainly used to feed the livestock that end upon as protein our tables. A good discussion of this can be found in the first section of Michael Pollan's *The Omnivore's Dilemma* [194]. Modern high-yield corn cannot be grown by hand. It requires intensive cultivation using industrial fertilizers and industrial harvesting. Also, most of it has to be transported several hundred miles to the feedlots. Most food in the United States travels a long way to reach our tables: it's been estimated that for every Kcal of food energy we consume 7–10 Kcal were spent bringing it to us. This means lots of energy spent on food transportation. If cheap transportation goes away suddenly, our food supply disappears. Everybody starves.

However, the oil supply won't evaporate overnight. Instead, the demand for light, sweet crude will exceed its supply sometime in the future. What happens then? Presumably the price of oil goes up, maybe rapidly, such as the $4-plus per gallon prices people in the United States saw a few years ago. What happens after that is unclear. We haven't developed any good substitutes for gasoline for transportation. This isn't to say there aren't any but that we haven't developed them yet. Supply-side economists predict that once the price of oil begins to rise, people will rapidly develop substitutes and the world will move away from crude oil as an energy source. This is possible: for example, synthetic gasoline can be produced from coal using the Fisher-Tropsch process, which kept the Germans fighting for at least one year after they were surrounded by the Allies at the end World War II. America has vast coal reserves. It isn't a great backup plan, as the process pollutes like crazy. Other sources possibly include biofuels. Energy conservation also helps, although it's a lot harder than most people think. The issue isn't so much overall energy resources but the fact that there is no really good substitute for oil to use for energy to power transportation.

Perhaps the best short story every written about the oil crisis is a fantasy story called "Not Long before the End" by Larry Niven. It was published in 1969, before the oil crisis. It can be read as an allegory of the energy crisis of the 1970s, although calling it this demeans the story somewhat. The story is set several thousand years ago in a fairly standard sword-and-sorcery setting. However, the main character, a wizard known as the Warlock, discovers that mana, the source of magic, is a nonrenewable resource. Use of magic by thousands of wizards across the world is draining away the source of their powers, and magic is becoming scarcer and harder to perform. A speech from the Warlock has stuck with me: "The swordsmen, the damned stupid swordsmen, will win in the end until mankind found another way to bend nature to his will." Several other stories were set in the same world, including the novella *The Magic Goes Away*, the short story collection *The Magic May Return*, and the novels *The Burning City* and *The Burning Tower*. The best-known movies concerning a dystopian future after the oil runs out are the Mad Max series, *Mad Max*, *The Road Warrior*, and *Mad Max: Beyond Thunderdome*, which took place after a war brought on by fighting over dwindling energy resources.

Whether or not the oil runs out 30 years from now, it will run out eventually. One other thing: the amount of time doesn't depend very much on how much oil there is, for the rate at which the world uses fossil fuels is increasing roughly exponentially. If the demand for a quantity doubles every 10 years, say, then in 20 years you will need to supply four times as much; in 30 years, eight times as much; in 100 years, 1,000 times as much. In 200 years the demand for the quantity has increased to 1 million times the initial demand, and it just keeps on going. Put differently, let's say we initially think we have enough oil to last the world for the next 30 years. Then we discover we actually have twice the amount we initially thought. If the doubling time is 10 years, then this is only an extra 10 years, supply, not an extra 30 years' supply.

If there is four times the amount we thought, this buys an extra 20 years, not an extra 120 years. And so on. The time it will take to run out increases logarithmically (i.e., very slowly) with the increase in the initial quantity.

Of course, we don't have to restrict ourselves to only one option for ending the world. Global warming combined with rising fuel costs due to Hubbert's peak can lead to a synergistic situation in which the stresses from the combination of the two are greater than the sum of the parts. We live in a world where a lot of the smaller, poorer nations have nuclear weapons; in the last five years, India and Pakistan have come scarily close to war. Rising food and fuel costs resulting from the two factors listed above could certainly lead to war, and the ubiquity of nuclear arms means that a country's destructive capacity doesn't depend primarily on its size or wealth. Also, the world's great oil reserves are in areas of great political instability.

17.2.4 Longer-Term Survival

Beyond this? If we last through our own turmoil, Mother Earth is very capable of doing away with us. The Earth goes through climate cycles, possibly caused by variations in its axial tilt. These variations are small, but enough to cause 100,000-year cycles of glaciation followed by briefer interglacial periods. We live in an interglacial period right now. If our world civilization can survive for several thousand more years, past the heating brought on by the man-made greenhouse effect, then we will see a cooling period as the next glaciers appear. All of modern human civilization—the development of agriculture, domestication of animals, writing, towns and cities—has emerged since the last glaciers retreated. Can human civilization survive the next glacial period?

The last 50 years have demonstrated that humankind can affect global climate. Perhaps we can do it in a safe way as climate cools. The temperature difference between the glacial periods and the interglacials isn't much: maybe 6°C. This is about how much the IPCC predicts the Earth will warm up due to anthropogenic greenhouse gases under the worst-case scenario. We have observational evidence that humanity can warm the Earth; the unanswered question is whether we can do it safely.

It is an interesting moral question whether we really should tinker with the Earth's climate and ecology on such a large scale.

Beyond this, a bevy of disasters wait for us. Beyond the glaciation cycle is the prospect of Earth's climate becoming completely chaotic. The Moon stabilizes Earth's axial tilt, but in 100 million years the Moon will have receded far enough away that the stabilizing effect will vanish. Earth's axis will wander drunkenly, and the long-term variation in temperature may destroy all life. But worse is to come: the luminosity of the Sun is gradually increasing. In a billion years the Sun's output will increase by 10%, leading to a runaway greenhouse effect that will make Earth a worse hell than Venus. If we cannot somehow temper this, our long-term descendants will have to find a new home, create one, or simply move the Earth.

NOTES

1. A very readable guide to the IPCC findings is Michael Mann and Lee Kump's book *Dire Predictions: Understanding Global Warming* [157].

2. To be very specific, these estimates apply to the "light, sweet crude" oil which is the oil of choice for the transportation industry: it is the type most easily refined into gasoline. Other types, such as the oil found in the Canadian oil sands, require more refinement before they can be used as a transportation fuel and are costlier to process.

CHAPTER EIGHTEEN

WORLD-BUILDING

18.1 TERRAFORMING

The noted physicist Kip Thorne first defined the Sagan problem as one
that tested what the laws of physics would allow on a fundamental
level [236]. The question Thorne was considering was whether physics
allowed time travel and faster-than-light travel. This is a question that
clearly probes the ultimate laws of nature. In these final chapters we're
going to consider a series of Sagan problems, each progressively more
fundamental than the last, to see what might be possible for a sufficiently
advanced civilization. These discussions are based on the laws of physics
as they're known now. However, all of them are practical impossibilities
(to say the least) for human civilization as it is now, and possibly will be
forever.

This chapter is titled "World-Building" because that's exactly what
we're going to consider. A large fraction of science fiction deals with the
transformation, or even construction, of habitable worlds for settlement
by humanity. Varied reasons are given for doing this, but the underlying
reasons seem to be variants on "because they're there." There's already
a lot of literary or historical critique of the assumptions behind this
sort of colonization. The fundamental one is that it is an inappropriate

projection into the future of the history of European and American colonial expansion of the nineteenth century. Perhaps a better critique of such ventures is to look at them from an economics point of view, which relatively few people have done so far.

We've already discussed the idea of building space stations to house humanity in low Earth orbit and space travel to other planets. This is distinct from the idea which we are considering in this chapter, actually "building" a habitable planet. In this context, "building" is a misnomer: in reality, we are considering the process of *terraforming* a world, that is, turning a planet such as Mars or Venus, on which people cannot survive in the open, into one in which people can survive in the open. Hence the name: we are forming other planets into ones like Earth.

Let's consider the preconditions needed for a planet to be ter-raformed. First, it must have a solid surface, which rules out the Jovian planets Jupiter, Saturn, Uranus, and Neptune. Second, it must be neither too hot nor too cold to support life. As we've seen in chapter 2, this is determined by two things: the distance of the planet from the Sun and the composition and density of the planetary atmosphere. The first condition is something we don't have any control over right now. However, we can change the composition of a planetary atmosphere. Since the Industrial Revolution we have been running an uncontrolled experiment on Earth's atmosphere to see what the results of changing the net atmospheric composition of CO_2 by a factor of two. So there are very good indications that if we can set up an industrial project of large enough scale on another world, we can change the climate of that planet significantly.

Orbital considerations leave only two planets worth considering, Mars and Venus. They are the only two planets close enough to the zone of life in the Solar System. They also have solid surfaces, unlike the outer gas giants. Science fiction writers have also considered the Jovian moons as targets for terraforming. In particular, Robert Heinlein wrote of a large-scale project to make Ganymede, Jupiter's largest moon, habitable in his excellent juvenile novel, *Farmer in the Sky* [111]. Although they are well outside the zone of life, Jupiter generates heat because it is slowly collapsing owing to its gravitational self-attraction [130]. As a result the moons Europa and Callisto have liquid water beneath their ice crusts, making them good candidates for life elsewhere in the Solar System.

The biggest issue is their lack of an atmosphere because of their low mass (roughly the same as that of Earth's Moon), and, ironically, their relative warmth, which tends to make planets lose their atmosphere more quickly. Titan, Saturn's largest moon, has an atmosphere because Saturn is farther from the Sun and hence much colder. Of course, science fiction speculation being what it is, Arthur Clarke in *2010: Odyssey Two* imagined the terraforming of Europa by aliens following the forced gravitational collapse of Jupiter, which transformed it into a small star [57].

In this chapter we will consider the terraforming of Mars specifically. This seems to be the easiest planet to consider for several reasons:

- It is essentially within the habitable zone.
- While it is very cold most of the time, its eccentric orbit ensures that at least for part of the Martian year, it warms up enough for liquid water to flow on its surface.
- It is close enough to Earth that we can imagine creating a more or less permanent settlement there.

Why not Venus? The crushing pressure and the extreme heat on the surface of Venus ensure that no one could live there or work there. The terraforming of Venus would be much more difficult.

18.2 CHARACTERISTICS OF MARS

Mars orbits the sun at an average distance of 1.52 AU. Its orbital period can be found from Kepler's third law:

$$T = (1.52)^{3/2} = 1.88 \, \text{yr}. \tag{18.1}$$

It spins on its axis with a period of 24 hours, 37 minutes, coincidentally much like Earth's orbital period, and the inclination of its rotation axis is about 24 degrees, again much like Earth's.

The atmosphere is about 95% CO_2, with trace amounts of argon and other gases, but is very thin. Atmospheric pressure at the surface is less

Table 18.1
Relative Properties of Earth and Mars

Property	Earth	Mars
R	1	0.533
M	1	0.107
T	1,000	140
V_e (m/s)	11,000	4,800
v_{rms} for H_2	3,500	1,860
He	2,480	930
O_2	878	328

than 1% Earth's sea-level pressure. There is essentially no free oxygen (O_2). As we saw in an earlier chapter, life on Earth is responsible for the large amount of O_2 in Earth's atmosphere: if the respiration cycle didn't replenish it continuously, it would react to form oxygen compounds. With no life (or very little, at least) on Mars, the life-giving components of the atmosphere are missing. As one writer described the Martian atmosphere, "it's almost too thin to be considered poisonous." The big issue with terraforming Mars is putting enough O_2 into its atmosphere for it to be breathable.

Table 18.2 shows some characteristics of Mars that will be useful in our discussion.

The amount of water on Mars is unknown, but certainly less than on Earth. As discussed in chapter 2, at one point Mars had enough atmosphere to warm it up enough for water to flow freely on its surface [130].

18.3   TEMPERATURE AND THE MARTIAN ATMOSPHERE

The two things needed to turn Mars into a habitable world are (1) increasing the average temperature to above the freezing point of water and (2) increasing and maintaining the O_2 content of the atmosphere to breathable levels. The high reactivity of oxygen means that to keep the

Table 18.2
Characteristics of Mars

Property	Value
Orbital distance (AU)	1.52
Orbital period (years)	1.88
Typical surface temperature (°C)	20 (day, equator) to −150 (night, poles)
Atmospheric composition:	
CO_2 (%)	95
O_2 (%)	0.13
N_2 (%)	2.7
H_2O (%)	0.002
Solar constant at Mar's orbit (W/m^2)	560

Source: Hester et al. [130, p. 218].

O_2 content at sufficient levels, the respiration cycle on a planetary scale must be created. The most commonly suggested means of doing so is by releasing genetically engineered extremophilic bacteria or plants into the Martian environment. "Extremophilic" refers to life that can survive in Earth's most challenging enviroments: in extreme heat or cold, under extremely high or low pressures, or in environments toxic to most other living creatures.

Increasing the temperature can be done in only one feasible way, by increasing the concentration of greenhouse gases in Mars's atmosphere. Again, this was once true of Mars, and maybe could become true again. There is currently only one important greenhouse gas in the Martian atmosphere, CO_2; any scheme to increase the average Martian temperature must begin by increasing its concentration. How much CO_2 do we need to generate? This is a tricky question to answer, but we can get an approximation by considering that the greenhouse warming of Mars by the CO_2 content in its current atmosphere is about 5 K [130]. If we assume that (1) the infrared absorption of the Martian atmosphere is proportional to the partial pressure of CO_2 in it; (2) it is independent of the other atmospheric components; and (3) there are no feedback effects once the concentration begins to rise, then we can calculate what we need.

The following is a back-of-the-envelope analysis. We begin with the observation that the temperature is not directly proportional to the atmospheric absorption but instead to its *fourth root* (see chapter 14). This means that a 1% increase in atmospheric absorption leads to a 0.25% increase in temperature. To make life simple, I'll assume that we need to increase the mean surface temperature by 80 K (i.e., increase greenhouse warming by a factor of 20), meaning that the partial pressure of CO_2 in the Martian atmosphere needs to increase by a factor four times this, or about 80. Since the partial pressure is currently about 6 mbar, it needs to increase to about 480 mbar.

Unfortunately, the situation is not nearly that simple. The absorption due to the infrared bands of CO_2 depends on the total atmospheric pressure because of an effect called pressure broadening, and on the column height of the atmosphere, which is higher on Mars than on Earth. Column height also decreases as the surface temperature increases, which means that as we change the Martian temperature, the total absorption will change in a complicated way . In addition, the amount of absorption doesn't depend simply on the partial pressure but exponentially on it. The simple approximation used above (percent temperature change is one-fourth the percent absorption change) is no longer valid when dealing with a very large temperature change. However, our simple analysis isn't too far off: Robert M. Zubrin and Christopher P. McKay have a sophisticated model indicating that a partial pressure of 800 mbar will lead to sufficient heating that liquid water can exist for long periods of the Martian year. Their model includes trapping of CO_2 by the Martian regolith, the rocky surface layer of the soil [256].

Most plans to terraform Mars begin by using methods to melt the Martian polar caps, which are almost entirely composed of frozen CO_2. Here, most of the schemes involve lowering the albedo of the ice caps by placing a layer of some dark material over them and letting the Sun do the rest: for example, a layer of soot has the advantage that the carbon needed to create the soot is already there. Carl Sagan was the first person to suggest doing this, in a paper published in 1973. He estimated that one would need to deposit about 10^8 metric tons of "a low albedo material such as carbon black" to create runaway melting of the polar caps [208]. Later, Robert Mole speculated that four nuclear bombs exploded over the Martian ice caps would generate enough sooty material to do the trick [167]. It is unknown whether there is enough CO_2 in the ice caps to increase the temperature sufficiently, although again it is clear than in the past there was certainly enough atmosphere for this. Unfortunately, this much CO_2 in the atmosphere is poisonous; indeed, partial pressures above 10 mbar are poisonous. Therefore, most schemes to terraform Mars hinge on the use of other greenhouse gases such as methane or water vapor. The trickiest part is that because Mars is farther from the Sun, you need a correspondingly larger amount of such gases in the atmosphere to warm it, which is a daunting task. In particular, the amount of water on Mars is almost certainly not enough to provide enough water vapor in the atmosphere for the needed change in temperature.

In the next section I consider the basic physics, chemistry, and estimated costs of producing a breathable atmosphere on Mars. However, before leaving the basic issues involved, let me direct the reader to a few more references that maybe useful if one is writing a story on this subject. Terraforming a planet is an unbelievably complex subject; Martyn Fogg's book is an indispensable resource for those interested in it [86]. As is appropriate, the largest section of the book is devoted to the case of Mars, which is the most plausible planet of choice in the Solar System. Fogg goes into much more detail than I can here and considers feedback effects, geography and geology, and other tricky issues. The book is unfortunately out of print and hard to find. It should also be considered out of date, given the massive amount we have learned about Mars in the past 17 years. Nevertheles, the book is a clear starting point for those interested in the subject.

A more sophisticated approach to warming of planets by greenhouse gases can be found in a paper by Akira Tomizuka, "Estimation of the Power of Greenhouse Gases on the Basis of Absorption Spectra" [238]. The subject of the paper is the greenhouse warming of the Earth, but it could be used to model the temperature of Mars as well. The study used a sophisticated mathematical/computer model of radiation transport in the atmosphere. It also used high-resolution absorption spectra of greenhouse gases, which most readers probably don't have access to. Zubrin and McKay give a simpler model for the surface temperature of Mars as a function of partial pressure and the solar constant [256, Eq. 1]. However, it should be treated as an approximation of limited validity only.

18.4 ATMOSPHERIC OXYGEN

The second part of the task of terraforming Mars is to generate oxygen. Most of the oxygen on Mars is bound up in the soil, in the form of rust (Fe_2O_3) or similar compounds, giving the characteristic red color of the planet. Generating oxygen to breathe means liberating the oxygen from the soil or other places, which will be a time-consuming task. The issue is that of energy: to free the oxygen from the soil, we need to run the following reaction:

$$2Fe_2O_3 + 1482\,kJ \rightarrow 4Fe + 3O_2.$$

For every mole of O_2 produced we need to expend a minimum of 494 kJ of energy. Another means of doing this would be through photosynthesis: if we could somehow induce the growth of some sort of vegetation on Mars it would liberate O_2 from CO_2 through the reaction

$$6CO_2 + 6H_2O + 2870\,kJ \rightarrow C_6H_{12}O_6 + 6O_2. \tag{18.2}$$

This was discussed in detail in chapter 4, where we applied this to growing food and producing the atmosphere for a space station. Photosynthetic energy would be free from the Sun (although it would also require water for the cycle, something of which Mars is in short supply), while the energy to liberate oxygen from the first process would be supplied in some sort of industrial process. One thing is clear, however: some sort of stable respiration cycle would have to be created at some point, or else the newly generated oxygen would simply chemically bond with the Martian "soil" again.

Either process involves a *lot* of energy. For example, photosynthesis requires 4.77×10^5 J per mole of O_2 produced, although plants are not able to utilize all of the energy which falls on them in the form of sunlight.

In fact, typical efficiencies are only of order $5 \times 10^{-2}\%$, although some plants can reach efficiencies as high as 1%.

We can do a back-of-the-envelope calculation to estimate the amount of energy it will require to create a breathable atmosphere on Mars. Atmospheric pressure at sea level on Earth is $10^5 \, \text{N/m}^2$, of which 25% is oxygen. Assume that we need a mix of roughly one-tenth this pressure as a minimum for the Martian atmosphere, or $2.5 \times 10^3 \, \text{N/m}^2$. This means that we need to generate approximately 600 kg of oxygen for every square meter of Martian soil. The molecular weight of oxygen is 32 g/mol, or $\approx .03$ kg/mol, so this represents $600 \, \text{kg}/.03 \, \text{kg/mol} = 2 \times 10^4$ moles of oxygen per square meter.

The radius of Mars is only one-third that of Earth, meaning that the surface area is about one-ninth that of Earth, or very roughly $10^{13} \, \text{m}^2$. For this area we need to generate approximately 2×10^{17} moles of oxygen. Through photosynthesis this represents a minimum energy cost of about 10^{23} J. The energy costs will be similar for any other chemical means used to generate oxygen. Because photosynthesis is very inefficient, we can assume that the true energy cost will likely be two to three orders of magnitude higher, or of order 10^{25}–10^{26} J. Currently, the world uses approximately 10^{20} J per year, so this represents about 10^5–10^6 years worth of energy use by the entire world. This is clearly an impractical project given current technology and energy resources.

The timescale is likewise daunting. The solar constant at the Martian orbit is 562 W/m^2; we have to divide this by four for latitude and day/night variation, so on average each square meter of Martian soil receives 140 W from the Sun. This represents a total insolation of $140 \, \text{W/m}^2 \times 10^{13} \, \text{m}^2 \approx 10^{15}$ W. Because it will take about 10^{25}–10^{26} J to generate this much oxygen via photosynthesis, it will take a time of about 10^{10}–10^{11} s, or 300 to 3,000 years to generate the needed oxygen. This assumes that we can use all of the sunlight which Mars receives, which is doubtful.

This shows the utility of using energy methods for quick feasibility calculations. It's not a detailed study of the issue, but by making an estimate of the available energy for this system we can state that Mars will not be terraformed in less than about 300 years even if we started today, which we clearly can't do. My feeling is that the timescale is likely to be much longer, possibly closer to 30,000 years or something like six times the length of recorded human history.

18.5 ECONOMICS

Another way to view the problem is by considering the cost to terraform Mars. As mentioned above, the minimum required energy is 10^{25} J, or 3×10^{18} kW-hr. (I'm assuming that even if we use sunlight via photosynthesis to power this, we will still have to pay for the energy somehow.) Current energy costs are about $0.1 per kW-hr; if we assume that in the energy-rich future they drop by three orders of magnitude, this represents a total cost of 10^{-4} \$/kW-hr $\times 3 \times 10^{18}$ kW-hr $\approx \$3 \times 10^{14}$. This is far higher than the current world GDP, and also rests on the assumption that energy costs are going to drop significantly. This is spread over a long time; we'll be liberal, and estimate that the time to terraform is a mere 1,000 years, meaning an investment of $\$10^{11}$ per year. This amount (\$100 billion per year) is less than the United States spends on its military but more than the entire basic science budget.

If we were to look to private industry to pay for this, we need to consider one fundamental issue: how is terraforming Mars going to turn a profit? Damned if I know. . . . There is nothing, and I mean *nothing*, that we could manufacture on Mars and ship to Earth that would not be infinitely cheaper to produce here. Anything sent here has to travel across 40 million miles of empty space. This is the weak point of most stories that involve terraforming other planets: the motivation for it is completely lacking. For example, the main motivation given to terraform Ganymede in *Farmer in the Sky* is to supply food to an overcrowded Earth. This is clearly nonsensical, and (in fact) in one part of the story, one character even does a calculation to show how absurd it is to try to ship items in bulk between the two worlds. The issue at hand was shipping soil from Earth to Ganymede, but the calculations hold equally well for shipping food from Ganymede to Earth. Handling the surplus population is again a loser as an issue: you can't ship people out fast enough. Maybe we can use the new world to hold our prisoners, as the British did in the eighteenth century, making Mars a new Botany Bay?

Why do I say that we're going to have to pay for this energy somehow? A lot of writers have written about terraforming a planet as if it were as simple as dropping a load of genetically engineered plants onto the planet and standing back as they grow and generate a breathable atmosphere. It isn't that easy. As one can see from the discussion here, the atmospheric constituents have to be carefully managed in order to heat the planet up and create the correct mix of a breathable atmosphere. We are barely beginning to understand the processes on Earth that maintain the temperature and atmospheric mix; granted, because Mars will be starting out from nearly a blank slate, these issues will (probably) be simpler than they are on Earth, *but humans will have to manage every aspect of them from the ground up.* This is a daunting task. Of course, some writers have taken the idea of world-building quite literally. This is the subject of the next chapter.

CHAPTER NINETEEN

DYSON SPHERES AND RINGWORLDS

For all the richest and most powerful merchants life inevitably became rather dull and niggly, and they began to imagine that this was therefore the fault of the worlds they'd settled on. None of them was entirely satisfactory: either the climate wasn't quite right in the later part of the afternoon, or the day was half an hour too long, or the sea was exactly the wrong shade of pink.

And thus was created the conditions for a staggering new form of specialist industry: custom-made luxury planet building.

—DOUGLAS ADAMS, *THE HITCHHIKER'S GUIDE TO THE GALAXY*

19.1 DYSON'S SPHERE

In June 1960 a paper titled "Search for Artificial Stellar Sources of Infrared Radiation" appeared in the journal *Science* [71]. Its author, Freeman Dyson, is a physicist who is currently a fellow of the Institute for Advanced Study at Princeton University (the same place where Einstein worked). We have encountered him before as one of the chief driving forces behind the Orion nuclear pulse propulsion drive. In the *Science* paper he presented an idea for finding alien civilizations that did not require searching for radio signals from them; instead, he suggested that astronomers look for sources of infrared radiation radiating at a temperature about right for life on a planet, but with the total power output of a star like our Sun.

> The material factors which ultimately limit the expansion of a technologically advanced species are the supply of materials and the supply of energy. At present the material resources being exploited by the human species are roughly limited to the biosphere of the Earth.... The quantities of matter and energy which might conceivably become accessible to us within the solar system are ... the mass of Jupiter ... [and] the total energy output of the sun [71].

Let us consider the twin issues of energy and resource needs versus their supply.

19.1.1 Energy and Resource Needs

Figure 19.1 shows an estimate of world population since the year 14 CE. It is taken from a 1977 publication, supplemented by more recent data [69, table 5]:

As one can see, world population increased dramatically between around 1800 CE and now, presumably owing to the Industrial Revolution. Not only did population increase dramatically, so too did the growth rate. From about 1800 to the present, world population has increased at an average annualized rate of about 1.3%. This doesn't sound like a lot, but I'll reintroduce a rule of estimation from an earlier chapter: the time it takes to double any quantity increasing at an annual rate of q% is $70/q$ years. This means that at current rates of expansion, the world population doubles every 54 years. Let's assume there are 100,000 habitable planets in our galaxy. If we could somehow shunt our excess population to these other (hypothetical) worlds, it would take only about 900 years until all of them had populations equal to that of present-day Earth.

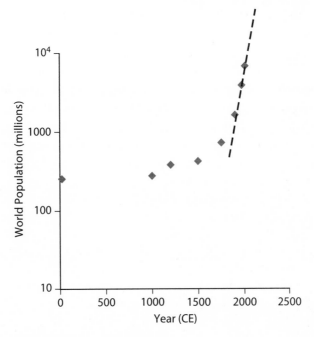

Figure 19.1. World Population over Time.

If we believe that faster-than-light travel is impossible, then this is clearly impossible. It would take far longer than 900 years simply to get to the other planets. Also, the energy requirements of placing that many people off-planet are probably impossible to meet. Freeman Dyson considered the problem of exponentially increasing population and energy demands in a 1961 paper. His question was whether one could figure out a solution to the population problem without interstellar travel.

19.2 THE DYSON NET

Dyson considered building numerous habitable space stations in orbit around the Sun at roughly the same distance as Earth is from the Sun. His idea for this solution came directly from science fiction. In interviews and published work he has credited Olaf Stapledon's book *Star Maker*, in which the members of an intergalactic civilization build large shells around their stars to capture the last remnants of energy as the stars die out [225]. Dyson's stations would occupy a large fraction of the total surface area of a sphere around the Sun of radius 1 AU, so that they would intercept most of the light coming from the Sun. It should be note that this network of stations is not a Dyson sphere; the Dyson sphere would be a *solid* shell built around the Sun that would intercept all the power coming from it. I'll refer to the multiple space stations design as a "Dyson net" to distinguish it from a Dyson sphere. The Dyson *sphere* has two major problems with it:

1. Newton's shell theorem states there is no net force between a solid spherical shell such as a Dyson sphere and the sun it surrounds. This means there is no way to prevent the shell from drifting off-center and hitting the Sun. There are even worse problems involving Larry Niven's Ringworld, which we look at in the next section.
2. Even though the net force is zero, there is still a stress on the sphere that tends to compress any given section of it. This stress is much higher than any known material can support. Again, we will consider this in detail for the Ringworld in the next section.

Large, individual space stations in orbit around the Sun don't have this problem. Each of them is in orbit, meaning that centrifugal force balances out the force of gravity acting on them. The tricky part is preventing the hundreds of thousands of space stations in orbit around the Sun from colliding with each other.

The amount of matter needed to build this shell of space stations is enormous. As Dyson stated, if we limit ourselves to material in the Solar System, most of the mass (outside the Sun) is concentrated in the giant planet Jupiter. Jupiter's mass is larger than all of the other planets' masses combined. It has a mass of 1.9×10^{27} kg, or roughly 318 times the mass of the Earth. Between Jupiter and Saturn we could probably scrape together about 400 M_E total, or about 2.5×10^{27} kg of material. That is a lot, but if we spread it in a thin shell with radius 1 AU around the Sun, it would have a thickness of only about 9 meters if the density were the same as water. However, this is more than we need.

19.2.1 Large Structures in Space

In a later paper published in 1996, Dyson considered the engineering issues associated with building large space-based habitats for the Dyson net [74]. He examined three questions in the paper:

1. Is it possible to build large structures in space?
2. Is it possible to build light, rigid structures in space?
3. Is it possible to take planets apart for the material needed to build these structures?

The constraints on building large structures in space are different from those that apply to building large structures on the surface of a planet. The ultimate constraint on how large a space station can be made is *tidal forces*. In some sense we've encountered this constraint already in considering the space elevator, although that structure represented a rather extreme version of the issue.

If we imagine a space station in orbit around a planet or star, there is only one point, the center of gravity, where centrifugal force balances out the attraction of gravity. Any part of the station "below" the center of mass, closer to the primary body, will see gravity as slightly larger than centrifugal force, and anything "above" it will find centrifugal force stronger than gravity. This means that there will be a seeming force trying to pull the station apart. This is what is known as a tidal force, The first time I heard of it was in Larry Niven's short story "Neutron Star," in which the hero of the story is nearly torn apart by these tidal forces when his spacecraft tries to get too close to said star [176].

The net stress pulling apart a structure of size L in a circular orbit of radius r around a primary body of mass M is of order

$$T \approx \frac{GMm}{r^3}L \approx (gm) \times \frac{L}{r}, \qquad (19.1)$$

where $g = GM/r^2$ is the acceleration of gravity owing to the primary and m is the mass of the space station. This is an approximation only; the stress will be a function of the shape of the craft and its distribution of mass. If the station has mean density ρ, then $m \approx \rho L^3$. Working this out, we find that

$$L_{max} \approx \left(\frac{Y_{max}r}{\rho g}\right)^{1/2} = (hr)^{1/2}, \qquad (19.2)$$

where Y_{max} is the maximum possible stress the structure can take and h is the parameter defined in chapter 5 for the space elevator: the maximum height a material of a given stress/weight ratio can stand in a gravitational field. One big difference here is that the gravitational field is that due to the Sun rather than the Earth. Putting in numbers appropriate for steel, I get a maximum size of about 9×10^8 m, or nearly 10^6 km, which is what Dyson found in his paper as well.[1] This is a structure roughly 150 times larger than the Earth. Another way to look at it is that the structure is larger than the distance of the Moon from the Earth. Large as that is, carbon nanotube fibers, with higher strength/weight ratios by a factor of 1,000 or more, would allow structures more than thirty times larger to be built.

The distance from Earth to the Sun is 1.5×10^{11} m. If we wanted enough of these structures to intercept all the light from the Sun, the approximate number is

$$N \approx 4\pi (r/L)^2 \approx 3 \times 10^5.$$

19.2.2 Building Light Rigid Structures

The question Dyson addressed next was whether it was possible to build light, rigid structures using the least amount of material. For structures of constant density, the mass of a structure is related to its size by the relation

$$M \sim L^3.$$

Because of the cubic dependence on size, mass increases rapidly as we build very large structures. Dyson considered what we would call today a "fractal," or self-similar, structure. This is built of subunits of octahedrons built into larger octahedral units, which are built into even larger octahedral units, and so on. Because of how the structure is built, and because it doesn't have to support its own weight when in orbit, one can build very large, very light structures. The details are in Dyson's paper [70]. Dyson estimated that building a structure 10^6 km in size would require a mass of about 3×10^{14} kg, or about $5 \times 10^{-11} M_E$.

Because of its fractal construction, the mass scales as a lower power of the size:

$$M \sim L^{3/2}$$

meaning that the overall density decreases as the structure gets larger. To build a Dyson shell one would need a total mass of about 10^5 times the mass of one of these huge space stations, or about 10^{20} kg.

This is still a lot of mass. The total mass of the Empire State Building is about 4×10^8 kg, so the Dyson net represents about a trillion copies of it put into space. This is a mass much larger than that of all the buildings in all the cities in the world put together.

19.2.3 Taking Planets Apart

The total mass required to build a Dyson net, while less than the mass of the Earth, is still huge by any standard. Could we imagine taking the Earth or another planet apart to rebuild it into these structures?

The first question to ask is how much energy it takes to deconstruct a planet. The gravitational potential energy of a uniform sphere is

$$U = -\frac{3}{5} \frac{G M_p^2}{R_p}. \tag{19.3}$$

A planet isn't a uniform sphere, but we'll take this as a starting point. In the formula, M_p is the planet's mass and R_p is its radius. Putting in values appropriate for Earth, we get $U = -3.8 \times 10^{31}$ J. In normalized units, we can write this (for any planet) as

$$U = -3.8 \times 10^{31} \times \frac{M^2}{R}, \tag{19.4}$$

in units where M is the mass of the planet relative to Earth's mass and R is the radius relative to Earth's. At least this much energy must be supplied to break up the planet. The luminosity of the sun is 3.86×10^{26} J/s; the energy it would take to disassemble Earth represents the total output from the Sun for a day.

This leads to the question of how to do it. Dyson imagined an ingenious scheme to increase the Earth's rotational speed until the planet flew apart. He proposed making an electrical motor out of the Earth. Laying large wires parallel to lines of latitude across it and running a current through them would give the Earth a sizable magnetic field. Then, running a current from pole to pole and out to large distances from the Earth and back again, one could generate a sizable torque either to speed it up or slow it down. Dyson envisioned doing this over the course of 40,000 years, which would require an average power of about 300 times the total power from the Sun intercepted by the Earth.

In Greg Bear's *Forge of God*, hostile aliens destroy Earth much more rapidly by dropping two large, ultradense masses into the center of the Earth. One is matter, the other antimatter, both at neutronium densities. When they merged together, the resulting explosion destroyed the world [32]. Based on the calculation above and using $E = Mc^2$, it would require a mass of about 10^{14} kg to do this. The purpose of this was destructive rather than constructive, but the net effect was the same: a planet in pieces.

19.2.4 Detection of a Dyson Net

Dysons claimed that his motivation for writing these two papers was not to suggest that humanity should actually build such a structure. To quote Dyson,

> When one discusses engineering projects on the grand scale, one can either think of what we, the human species, may do here in the future, or one can think of what extraterrestrial species, if they exist, may have already done elsewhere. To think about a grandiose future for humanity ... is to pursue idle dreams. ... But to think in a disciplined way about what we may be able to observe now astronomically ... is a serious and legitimate form of science [74].

He wanted to see if astronomers could detect such structures if they existed. The issue is that the power output from a star like our Sun follows a blackbody curve with a characteristic temperature of about 6,000 K. This means that the peak in the spectrum is at a wavelength of about 0.5 μm, or 5×10^{-7} m. However, the Dyson net's mean temperature would be a lot lower, around 300 K. The shell would absorb the sunlight at a relatively high temperature and reradiate it at a lower temperature. An astronomer would see a blackbody curve with a peak wavelength of about 10 μm, or 10^{-5} m. There are other astronomical sources of radiation at this wavelength. However, unlike other sources, it would look like a star with the same luminosity as our Sun but with a planet-like temperature. In fact, one can argue that if a civilization didn't completely surround the star, astronomers would see two blackbody curves of similar strength: one resulting from the partly blocked star itself at a high temperature, and one from the shell at a much lower temperature. This seems a reasonable and unique signature for such a structure. I say "reasonable" because it is probably impossible to block the star entirely. The Dyson net isn't a solid shell but a large number of relatively small satellites in orbit around it. To keep the satellites from colliding, there should be some room between them. To spot a Dyson net, one might look for stars like our Sun that radiate most of their energy in the visible region of the spectrum, but also a relatively large fraction (perhaps 10%–50%) in the far infrared. I can't think of any naturally occurring astronomical objects that have a spectrum like that, but perhaps there are some.

19.3 NIVEN'S RINGWORLD

There are obvious issues with building a Dyson shell: the energy it requires, the need to destroy a world to create it, and so on. In 1970 Larry Niven came up with a smaller structure than a Dyson sphere for his novel *Ringworld* [177]. As the name implies, the Ringworld is a ring around a star, of radius 1 AU and width to be determined. Niven decided to spin it to provide gravity (in the form of centrifugal force) for its inhabitants.

What I'd like to do in this section is to play "Ringworld engineer"— that is, go through the process that (presumably) Larry Niven did when originally designing the thing. It's a fun exercise, and the concept reflects the best of what hard science fiction has to offer; of course, our job here is much easier than Niven's was: we have only to reconstruct the idea, not come up with the original notion. One other thing: I am deliberately not referring back to the book for the parameters but will estimate them based on what I know of physics and astronomy. At the end, we'll compare them to the ones found in the book.

19.3.1 Ringworld Mass

First, let's estimate its mass. It is considerably larger than a planet but smaller than a star. Unlike the Dyson net, it is a rigid structure. Oddly enough, its mass will be larger than what we calculated for the Dyson shell.

It's larger than a planet but smaller than a star. Perhaps we can use the geometric mean of the mass of Earth and the Sun for its mass. Since the Earth's mass is roughly 10^{-6} of the Sun's mass, the geometric mean of the two masses is roughly 1/1,000 the Sun's mass, or 1,000 times the mass of the Earth: about 6×10^{27} kg. Another way to estimate the mass is to try to figure out where all this mass should come from. If we assume the designers used only the resources available in one stellar system, then we should estimate the Ringworld mass as the mass of everything in the Solar System but the Sun. As I stated above, most of the mass in the Solar System apart from the Sun is in the planet Jupiter. The combined mass of Jupiter and Saturn is roughly 390 M_E, which is not too far from my first estimate. Of course, astronomers have found "super-Jupiters" in other stellar systems with masses more than ten times Jupiter's mass, so there are good reasons to think that I can be liberal by a factor of two or even more. I will use an estimate of 5×10^{27} kg to make the calculations ahead come out nicer.

19.3.2 Ringworld Radius and Mean Temperature

One parameter of importance is the radius of this ring around the Sun. Because the Ringworld is inhabited by humans and humanoid aliens such as Kzinti, the distance from its star must be *about* the same as the distance from Earth to the Sun. If not, the radiant flux from the star will make the world too hot or too cold. On this assumption, $R \approx 1 \text{ AU} = 1.5 \times 10^{11}$ m. There is one caveat: given this distance, one might naively expect the Ringworld temperature to about the same as Earth's, but this is ignoring the fact that greenhouse warming by Earth's atmosphere leads to a significant rise in Earth's mean temperature (roughly $30\,^{\circ}$C). "So what?," I hear the fan say: "Ringworld's atmosphere, just about the same composition as Earth's, leads to greenhouse warming of the Ringworld as well." Yes, but Earth's atmosphere extends completely around the planet, whereas Ringworld's atmosphere is only on the inner side. One must apply the principle of detailed balance to calculate the temperature, using the fact that part of the heat flux escaping from the structure leaves through the back. Let's calculate the mean temperature of the Ringworld using the principle of detailed balance.

The basic principle of calculating the mean temperature of this structure is the same as calculating the mean temperature of a planet: there is a certain flux absorbed on average from the Sun by the structure, which depends on the total flux from the star and the mean albedo. It is radiated away in the form of infrared radiation; however, some of the infrared radiation is trapped by the atmosphere, leading to an increase in the mean temperature. There are two complications: first, as mentioned above, half of the flux is radiated away by the back of the structure where there is no atmosphere. All other things being equal, this will lead to a net cooling effect. Second, day/night alternation and latitude variations in mean insolation for a planetary surface lead to a reduction in the effective flux by a factor of four. Control of solar flux heating the Ringworld is by means of "shadow squares" in orbit around the Sun that periodically interrupt the sunlight to effectively give variations between day and night. The godlike engineers who construct the structure will choose the spacing and width of the shadow squares to maintain the proper temperature, which we will assume to be 288 K, the same as for Earth. (Another means of controlling temperature is via atmospheric composition, that is, by changing the mix of greenhouse gases in the atmosphere.) A little algebra gives us

$$T_{RW} = \left[\frac{\eta F}{\sigma(2 - f/2)} \right]^{1/4}. \tag{19.5}$$

In this expression, T_{RW} is the mean temperature of the structure, F is the mean "solar" flux illuminating the surface (assumed to be 1000 W/m^2), σ is the Stefan-Boltzmann constant (5.67×10^{-8} W/m$^2 K^4$), f is the fraction of emitted infrared radiation reabsorbed by the atmosphere (0.77, as for our model of Earth's atmosphere), and η is the fraction of time the Ringworld is sunlit; it is our free parameter. The Ringworld is made up of a particular brand of unobtainium called *scrith*. I am assuming that *scrith* perfectly absorbs all light incident on it; that is, it has an albedo of zero. Playing around with the model leads to $\eta = 0.65$ for a temperature $T_{RW} = 288$ K. This means that the shadow squares should be set up in such a way that 65% of the time, the Ringworld is in daylight and 35% of the time it is night.

19.3.3 "Gravity" and Rotational Velocity

Because of the small thickness of the structure (discussed below), its gravitational attraction will be completely insufficient to hold anything to it. Because of this, Niven posited the structure was orbiting with such a velocity that the centrifugal force effectively served the purpose of gravity, exactly as in a rotating space station. From this, and the assumption that the acceleration of gravity is effectively the same as on Earth ($g \approx 10$ m/s^2), we can calculate the rotational speed of the structure:

$$v = \sqrt{gR} = \sqrt{10 \, \text{m/s}^2 \times 1.5 \times 10^{11} \, \text{m}} = 1.22 \times 10^6 \, \text{m/s}. \qquad (19.6)$$

This is forty times the orbital velocity of Earth; it also means that the kinetic energy of the structure will be humungous:

$$K = \frac{1}{2}mv^2 = 3.65 \times 10^{39} \, \text{J}. \qquad (19.7)$$

Pay close attention to this number.

Moving on: Let's say, as a wild guess, that *scrith*'s density is typical for solid matter: about 5,000 kg/m^3. Then the total volume of the Ringworld is 10^{24} m^3. We can view the Ringworld as a thin ribbon with radius R, thickness T, and width W. The volume is given by the formula

$$V = 2\pi RWT. \qquad (19.8)$$

Only R is specified initially; for Earthlike life to flourish (assuming the star is similar to the Sun), $R \approx 1 \text{ AU} = 1.5 \times 10^{11}$ m. Since the Ringworld is nominally created to handle overpopulation, the inner surface area should be as large as possible, meaning we want to make the width as large as possible, or the thickness as small as possible. As a wild guess, I'm going to assume (similar to what we did for the total mass) that the width is the geometrical mean between the radius and the thickness:

$$W = \sqrt{RT}.$$

From this guess, and using equation 19.8, we arrive at an equation for the thickness:

$$T = \left(\frac{V}{2\pi R^{3/2}} \right)^{2/3} = 19.6 \text{ km.} \qquad (19.9)$$

This leads to a width of 5,400 km, which is roughly half the Earth's diameter, and a surface area given by

$$A = 2\pi WR = 5.1 \times 10^{19} \text{ m}^2,$$

or roughly 10^5 times the surface area of the Earth. Looking up what Niven wrote, he assumed a diameter of about 1 million miles, or 1.6×10^6 km, meaning a surface area of about 1.5×10^{21} m^2. This is about three million times the surface area of Earth [177]. However, it means the structure is now only about 700 m thick.

19.3.4 Ringworld Structural Strength

Larry Niven once wrote that the Ringworld could be understood as a suspension bridge with no endpoints [180]. This description is worth examining in some detail. First, like the cables in a suspension bridge,

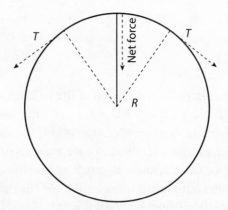

Figure 19.2. Ringworld Structural Tension.

the structure is entirely in tension. This is because the structure is in such rapid rotation that the centrifugal force pushing out on it is *much* greater than the attraction of the star's gravity on it. Planets are different: a planet's orbit is essentially defined by the balancing point of centrifugal force and gravity, so (apart from tidal forces) one doesn't need to worry about these considerations. In the rotating coordinate frame of the structure, the centrifugal force pushing outward on the ring is balanced by the net component of the tension in the ring pointing inward. Figure 19.2 illustrates this: the net force per unit circumference acting on the element of the ring shown is twice the tension multiplied by the ring curvature ($= 1/2R$).

$$\rho T W g_{eff} = \frac{\tau}{2R} \qquad (19.10)$$

or

$$\tau = 2 g_{eff} \rho T W R = M g_{eff}/\pi = 1.6 \times 10^{28} \, \text{N}.$$

Here, g_{eff} is the effective acceleration of gravity on the structure ($= v^2/R$), which is assumed to be $10 \, \text{m/s}^2$, M is the total mass, and ρ is the density of *scrith*. This has a nice interpretation: the net tension in the structure is the total centrifugal force on the structure divided by pi. The stress in the structure is the tension per unit cross-sectional area, which is again easy to work out:

$$\sigma = 2g_{eff}\rho R = 1.5 \times 10^{16}\,\text{N/m}^2.$$

The tension in the structure is some five orders of magnitude higher than the tension in the space elevator, and there are no materials that currently exist that we could build *that* structure with. In the "small favors" department, the stress doesn't depend on the structure's thickness, meaning that effectively we can build as thin as we like. One point, however, is that the bulk modulus of the structure should be at least an order of magnitude larger than the tension, or else the structure will begin to deform significantly: this implies a bulk modulus of about $Y = 10^{17}\,\text{N/m}^2$ at a bare minimum. However, the speed of sound (i.e., the speed of compressional waves) in a structure is given by the formula

$$v_s = \sqrt{Y/\rho} = \sqrt{10^{17}\,\text{N/m}^2/5{,}000\,\text{kg/m}^3} \approx 5 \times 10^6 \text{m/s}, \qquad (19.11)$$

or a whopping 1.5% of the speed of light!

Scrith almost certainly can't exist. Ordinary matter is held together by electrostatic forces; the maximum possible value for the bulk modulus is going to be of the order of the stored energy per unit volume for bulk matter, which is about $10^{12}\,\text{N/m}^2$. This makes the space elevator a marginally possible concern but the Ringworld an almost certain impossibility. *Scrith* is not quite as extreme a material as the *gleipnirsmalmi* needed for faster-than-light travel, but it's approaching it. The h parameter (assuming a density of $5{,}000\,\text{kg/m}^3$) is about 2×10^{12} m, or $\sim 10^{-3}$ light-years. *Scrith* isn't quite *gleipnirsmalmi*, but it isn't trivial to find either.

19.3.5 Energetics

Assuming that it is possible to build, how long would it take to build such a structure? This isn't an easy question to answer: the structure is so far beyond what humanity can do now, and there are so many assumptions one would have to make about the society that could build it, that, well, words fail me. However, maybe we can make a lower bound based on energy.

We calculated the kinetic energy of the structure: 3.65×10^{39} J. Somehow the civilization has to generate the energy to give it this rotational kinetic energy. The handiest source of energy is the star the Ringworld circles. If the star is similar to our Sun, its luminosity is about 3.6×10^{26} W. If we assume that this super-civilization can harness all the energy of the star, it will take a total time of about 10^{13} s, or 300,000 years, to generate all this energy from the star. However, it's probably unreasonable to assume the super-civilization can use *all* the energy from the star. If it harnesses only 10% of the energy, the figure goes up to three million years.

Is there another way to generate this energy? Well, Einstein tells us that the total energy content of matter is $E = Mc^2$; if we could extract the total energy content of matter to rotate this structure, the mass we would need to convert to energy is 3.6×10^{39} J$/ \left(3 \times 10^8 \, \text{m/s} \right)^2$ $= 4 \times 10^{22}$ kg. This is roughly the mass of Earth's Moon (7.35×10^{22} kg). However, you need some means of extracting this energy; rumbling about antimatter won't cut it, as it takes energy to make antimatter, as we saw in an earlier chapter. If the Ringworld engineers don't have a handy moon made of antimatter (which they might—one of Niven's *Known Space* stories has Beowulf Schaeffer investigating just such an object), they will need to extract the energy another way. The only reasonable means is to toss the object into a rapidly rotating black hole; one can extract up to 50% of the mass-equivalent energy by doing so (more on this later). Again, story continuity helps us here: another of the Beowulf Schaeffer stories involves him dealing with a "space pirate" who is using a miniature black hole to swallow starships. Of course, the black hole must be large enough that Hawking radiation hasn't caused it to evaporate, but again, more on this later.

19.4 THE RINGWORLD, GPS, AND EHRENFEST'S PARADOX

One neat thing about the Ringworld is that it is a living embodiment of one of the most puzzling features of Einsteinian relativity, as the physicist Paul Ehrenfest pointed out in 1909. The idea is simple: imagine that Louis Wu, the protagonist of the first three Ringworld novels, is in a spacecraft that is hovering directly over Ringworld's star.

We'll approximate the Ringworld as a perfect circular ring with radius of 1 AU, centered on the star. If the Ringworld were unmoving, then its circumference would be equal to 2π AU; however, since it is rotating, the theory of relativity predicts that it is foreshortened in the direction of its motion. Therefore, by that argument, its circumference should be less than 2π times its radius.

This is not a big effect. Even with the Ringworld's high rotational speed of 1,200 km/s, the effect is only about one part in 10^5. However, the Ringworld is huge: the amount "missing" from the circumference is more than half the size of the Earth! Another way to put this is that this relativistic effect is more than 100,000 times larger than relativistic effects induced by Earth's rotation, and those effects are easily measurable by atomic clocks. Indeed, it might be difficult to design a GPS-equivalent system for the Ringworld because of this huge correction.

Ehrenfest's paradox is notoriously difficult to handle. One can argue that the circumference should, in fact, really be larger than 2π AU— at least for some observers. Some would argue that Wu, because he is observing the ring from an inertial reference frame, should measure the circumference as 2π times its radius. However, let's look at what happens when someone on the Ringworld tries to measure the circumference. Let us say that Chmeee, a Kzinti and a friend of Louis Wu, has a large supply of measuring sticks exactly 1 m long. He is going to use them to measure the circumference of the Ringworld by laying them in place end to end. From Louis Wu's perspective, each measuring stick is foreshortened in the direction of motion, so it takes more of them to measure the circumference than if the Ringworld weren't rotating. Chmeee is going to measure the structure as being *longer* than $2\pi R$!

Most textbooks on relativity give short shrift to Ehrenfest's paradox, mostly because it doesn't have a clear-cut solution. One that has a decent discussion of the issue is *Relativistic Kinematics*, by Henri Arzeliès [28, pp. 204–243]. It devotes an entire chapter to the problem. To summarize a long and complicated argument, Arzeliès concludes that the problem is incomplete without recourse to the general theory of relativity. This is understandable: general relativity is a theory of curved space-time that allows for non-Euclidean geometries where the ratio of the circumference of a ring to its radius is not equal to 2π. His conclusion is that one must know the material properties of the ring, which then must be fed into Einstein's field equations to discuss its subsequent motion and shape.

Most writers don't agree with Arzeliès, although it is more a matter of interpretation than of fact. The consensus seems to be that initial conditions matter a lot for this problem. How you start the structure rotating is important in determining its geometry. The question is a fascinating one, however, and far from settled. A relativistically rotating Ringworld would be the ideal object to settle the debate once and for all.

19.5 THE RINGWORLD IS UNSTABLE!

Oh, the Ringworld is unstable,
The Ringworld is unstable
Did the best that he was able
And that's good enough for me!
—SCIENCE FICTION CONVENTION FILKSONG

Shortly after the publication of the original novel *Ringworld*, Larry Niven was greeted by chants from students at a lecture at MIT, "The Ringworld is unstable!" This means that, left to its own devices, the Ringworld has a tendency to slide into its sun in a fairly short time. And by "fairly short time," I mean a fairly short time on human timescales, not astronomical ones: at most, a few years. This instability drove the plot of the second novel, *The Ringworld Engineers*, as Niven was compelled to find a solution to it.

Interestingly enough, while the instability is pretty famous in science fiction circles, there aren't any complete descriptions of its causes. Larry Niven wrote in an essay that his friend Dan Alderson, a scientist at the Jet Propulsion Laboratory in Pasadena, spent two years working out the exact mechanism of the instability; unfortunately, as far as I can discover, he never published it. There have been at least two scientific papers written by others on the nature of the instability, but they are unconvincing; while they are correct as far as they go, they are fairly simplistic and ignore one or more of the complexities imposed by the nature of the structure. There are three issues at stake: a static instability, a dynamic instability, and a tendency for the structure to tear itself apart.

Before I begin my analysis, I want to mention three things: first, my assumption throughout this is that both Larry Niven and the late Dan Alderson are (were) very bright people, at least as bright as I am. Also, I don't have two years of leisure time to try to repeat his entire analysis. However, what one person can do, another can imitate, and I am bright enough (I think) to be able to repeat the key features. Second, I am standing on the shoulders of giants: what follows is based on a paper written by the third greatest physicist of all time, James Clerk Maxwell, on the stability of Saturn's rings [161]. Maxwell's paper is a tour-de-force analysis showing that Saturn's rings can't be solid or liquid in nature but must consist of small fragments, and forms the basis of the mathematical field of stability analysis as it exists today. The paper is key to understanding the Ringworld instability in its fullness, although the techniques must be adapted to the problem at hand, for reasons discussed below. Finally, the mathematics of such an analysis go beyond the level of mathematics used in this book; I will discuss it in only a general way. Also, the instability discussed relates only to in-plane motions of the structure, that is, motions of the ring in the plane that contains its sun. Out-of-plane motions aren't considered here at all.

19.5.1 Static Instability

This is the issue that has been written about. If the Sun is at the exact center of the ring, then the net force on the ring is zero because the gravitational attraction of the Sun on one part of the ring is exactly cancelled out by the force on a part exactly opposite it. However, if the ring slides a little off-center or is displaced slightly upward or downwards, what then? If it tends to come back to the original position, then the system is stable. Small perturbations of the structure tend to restore themselves. If, however, it tends to slide further off-center, then it is unstable, and the ring will be destroyed when it hits the Sun.

Colin McInnes did a stability analysis focused on this question [162]. He found that the Ringworld was unstable to motions in the plane. That is, if the Ringworld slides off to one side, gravitational forces tend to force it more to that side instead of back to the center. The analysis is straightforward, but the mathematics involves hypergeometric functions, so we will not reproduce it here. Out-of-plane motions are stable, however. If the structure is displaced upward, gravity tends to force it back downward, and vice versa.

This is not surprising. No *static* configuration of masses can be in equilibrium owing to gravitational forces alone. This is a consequence of what is called *Earnshaw's theorem*. A good paper on Earnshaw's theorem for the advanced reader is W. Jones's 1980 paper [133]. A planet orbiting the sun is in dynamic equilibrium: gravitational forces are balanced by "centrifugal" ones. This is why I think that the Ringworld instability is much more subtle than a mere static problem. Planetary motion is stable over timescales of billions of years, so why couldn't the Ringworld be stabilized the same way?

19.5.2 Dynamic Instability

The issue of the stability of solid ring systems goes back about 150 years. The issue did not concern man-made structures. Instead, the question was whether Saturn's rings were solid, liquid, or composed of many smaller bodies rotating around Saturn.

Before *Voyager 2* flew by Saturn we had no detailed pictures of the ring system, but it had long been understood that they couldn't be solid because of an analysis done by James Clerk Maxwell, one of the greatest physicists who ever lived. Maxwell is best known today for his discovery of the full set of equations governing the electromagnetic field, but he made contributions to all areas of physics. In 1856 he published an essay on the stability of Saturn's rings that included their motion [161]. He showed that if Saturn's rings were solid, they couldn't be stabilized by their rotation in the same way that a planet orbiting the Sun could be. They would inevitably slide off-center and hit the planet. Because they didn't do this, he concluded that the rings were made from small chunks of material orbiting the planet independently. He also considered whether they could be liquid; the answer was also no.

The stability issue is rather subtle. To begin with we need to look at why planetary orbits around the sun are stable. A planet, as it orbits its star, seems to feel two forces acting on it: the force of gravity pulling it in and a "centrifugal force" pushing it away. For a circular orbit, the two forces balance out at all times. For an elliptical orbit, as the planet approaches the star, the centrifugal force is stronger than the gravitational force, resulting in it being pushed away. As the planet gets farther from the star than its average distance, the force of gravity is stronger than the centrifugal force. This results in a net force toward the star. The final result is that the planetary orbit is stable: when the planet is too close, it is pushed away; when it is too far, it is pulled in.

This isn't true for a Ringworld structure. Here the centrifugal force pushing outward is much stronger than the gravitational force pulling inward. The total force is balanced out by one that doesn't exist for a planet: the tension in the ring structure. The Ringworld is a lot like a rapidly spinning gyroscope. If the structure slides off-center, it'll start to wobble. The wobble will couple into the motion of the ring, sliding it further off center, and soon. This will eventually crash the structure into the star, but it will not be a smooth motion at all. The wobble may get so strong that it tears the structure apart before it hits the sun.

19.5.3 Deformation

The final issue is deformation of the structure. If the ring deforms a little bit from a perfect circle, does the structure tend to "bounce back" to a circle or does it tend to collapse itself out into a line? I don't know the answer to this question. It probably depends on the material properties of *scrith*. Such stability problems tend to be highly nonlinear, and therefore difficult. My suspicion is that the extremely high centrifugal force will complicate the problem a lot.

19.5.4 Other Large Structures

In the recent novel *Bowl of Heaven*, Gregory Benford and Larry Niven write about another type of large structure. Essentially it is a half Dyson sphere. The structure looks like a parachute being dragged behind a racing car, with the Sun functioning as the car and the world-bowl as the parachute [37]. It is being used as some sort of very large starship. The structure was built for mysterious purposes; humans stumble across it during the course of the first interstellar exploration. One of the characters says of the structure:

> The shell should fall into the star—it's not orbiting. There's some sort of force balance at play. . . . Just spinning it isn't enough, either—the stress would vary with the curvature. You'd need internal support. [37, p. 33]

I have no idea whether Benford and Niven did a stability analysis for this structure. To give them credit, they mention in the book the misconception that Dyson's original idea was a solid structure:

> Only—the old texts reveal quite clearly that Dyson did not dream of a rigid structure at all. Rather, he imagined a spherical zone filled with orbiting habitats, enough of them to capture all of the radiant energy of a star. [37, p. 33]

In the 1970s Larry Niven wrote an essay "Bigger Than Worlds," in which he discussed different structures from the size of large space stations to the galaxy [180]. I think they are all implausible for reasons that were brought up when Dyson published his original paper.

19.6 GETTING THERE FROM HERE—AND DO WE NEED TO?

Dyson originally proposed these structures as a cure for overpopulation. However, Dyson's assumption was that the world population would continue to increase exponentially in a Malthusian fashion. Exponential increase is characteristic of populations that have plenty of resources. This can be seen in bacterial populations; given sufficient resources, they double every generation until the food available to them is exhausted. Human population has increased at an average rate of about 1% per year since around 1800 CE, that is, since the beginning of the Industrial Revolution [69]. However, this rate of increase was made possible by the enormous increase in resources available.

The growth of a bacterial or animal population doesn't follow an exponential curve indefinitely. It follows the same sigmoidal curve that we investigated when looking at the Hubbert peak for oil use. It is characteristic of an expanding use of a finite resource. Dyson's point was that the expansion of human population is limited by the resources, chiefly energy, available to it. The question is whether the human race will expand to this point if such resources are available to it. After Dyson's original article was published, John Maddox, Eugene Sloane, and Poul Anderson published rebuttals to his thesis in *Science* magazine [156]. Anderson made the rather cogent point that unrestrained population growth would make the structure nearly impossible to build because, at the growth rates postulated by Dyson, the structure would take much longer to build than the time it would be needed in. To quote Anderson:

> Even Dyson intimates that the project would take several thousand years to complete. In short, uncontrolled population growth will make the construction of artificial biospheres such as Dyson spheres impossible, and birth control will make them unnecessary. [156, p. 257]

It is a truism that First World civilizations have lower birth rates than Second or Third World ones. This is due to easy access to birth control and (probably) to other social factors, equality of rights for women being chief among them. If this trend holds for the developing world as in the developed, overall world birth rates would be expected to drop. (It's interesting to note that one predicts the same result if one assumes resource depletion and the Four Horsemen.) In this optimistic scenario, world population should level off or possibly even decrease in the future. In any event, birth control seems easier in the long run. However, this still doesn't rule out highly advanced civilizations.

Note

1. For the same weight/strength ratio, structures in Earth's orbit are limited to about 100 km in size because of the higher value of g and lower value of r. This is another reason why it is difficult to build a space elevator.

CHAPTER TWENTY

ADVANCED CIVILIZATIONS AND THE KARDASHEV SCALE

20.1 THE KARDASHEV SCALE

> Any sufficiently advanced technology is indistinguishable from magic.
> —ARTHUR C. CLARKE, *PROFILES OF THE FUTURE*

In the 1950s, Isaac Asimov envisioned a great "Galactic Empire" as the stage for his *Foundation* novels, a civilization set thousands of years in the future that encompassed every star in the Milky Way Galaxy. The sweep of this vision proved popular: Asimov followed his initial three novels, focused on the collapse of the first empire, with three more that both chronicled the rise of the second empire and also deconstructed the background of the first novels. Not only have several more *Foundation* novels been written posthumously (a neat trick, that), but his vision has been shared by many other science fiction writers since his time. Galactic civilizations have featured in the stories of Cordwainer Smith, with his "Instrumentality of Mankind"; in Larry Niven and Jerry Pournelle's Codominium and Empire of Man stories and novels, including *The Mote in God's Eye*; and in the various *Star Trek* series, novels, cartoons, and films and the *Star Wars* films and novels. The combination of Napoleonic era intrigue with ships popping in and out of hyperspace is almost irresistible.

We've devoted two previous chapters to discussing the problems inherent in interstellar travel. The issues of cost, time, and energy consumption effectively nix travel between the stars for a civilization like ours. However, our technological civilization is historically extremely young, and technological developments have shown explosive exponential growth. Maybe humanity can eventually develop some sort of galactic empire (or at least some kind of galactic civilization) if given sufficient time to grow and sufficient development.

In 1964, in a paper on the detection of advanced alien civilizations, Nikolai Kardashev, a Soviet astronomer, proposed a typology of advanced civilizations:

> Type I—A civilization that could harness the energy resources available to the entire planet.
>
> Type II—A civilization that could harness all the energy produced by its planet's star.
>
> Type III—A civilization that could harness all the energy produced by its galaxy.

Actually, in the original paper, a Type I civilization was defined as being like the technological civilization we have today; no mention was made of being able to harness all of the resources available to the planet. However, since the original publication, the term has taken on that meaning.

20.2 OUR TYPE 0.7 CIVILIZATION

World energy usage rate is currently about 10^{13} W. This is the average amount of energy used by everyone in the world every second of every day. The United States is the largest energy consumer, accounting for about 25% of the total amount, although China and India may surpass the U.S. share within the next half century. Carl Sagan took Kardashev's original loose classification and wrote out a mathematical formula to place the different civilizations into these categories:

$$K = \frac{\log_{10} P - 6}{10}.$$

(20.1)

Here, K is the Kardashev scale rating; the "$\log_{10}$" refers to taking the logarithm, base 10, of the average power used by the civilization. Usage of the logarithm compresses the scale; for example, $\log_{10} 10 = 1$, while $\log_{10} 100 = 2$. Because of the division by 10 in the formula, to increase by one unit on the Sagan-Kardashev scale, a civilization must increase its power usage by a factor of 10 billion. Putting in the numbers, our own current world civilization uses 10^{13} W; $\log_{10} 10^{13} = 13$, and $(13 - 6)/10 = 7/10 = 0.7$. Our civilization thus stands at 0.7 on the Kardashev scale.

I want to consider what it will take for us to become a Type I civilization, but let's first take a look at preindustrial civilization on the same scale. The big thing that separates pre- from postindustrial civilization is energy usage rates. Before the Industrial Revolution, people were limited in their power usage to the amount of power that a human body (or in select cases, draft animals) can exert—roughly 250 W. In a hypothetical civilization of 10 million people, this implies a power usage of roughly 2.5×10^9 W. The Kardashev rating can be found easily: $\log_{10} 2.5 \times 10^9 = 9.4$; $(9.4 - 6)/10 = 0.34$. This gives some idea of the relative ratings of the world civilizations; even though world civilizations are very different from one another, they occupy a fairly small range on the Sagan-Kardashev scale. We can reverse the formula to find the energy usage for a given value of K:

$$P = 10^{10K+6} \text{ W}.$$

(20.2)

A Type I civilization has an average power usage of $10^{10 \times 1+6}$ watts $= 10^{16}$ W, or 1,000 times the amount of power consumed by the world today.

If we guess that energy usage in the world increases at an average rate of about 3% per year, then energy usage doubles roughly every 23 years. If we take a number and double it, then double it again, then repeat that again, after ten doubling periods the original number will have increased by a factor of just about 1,000 (really, 1,024, but we're rounding off). So, ten doubling periods = 10 × 23 years = 230 years until we become a Type I civilization. This makes some sense: by the calculation I did earlier, we are about as far away from a Type I civilization as a preindustrial civilization is from us (i.e., a distance of about 0.3 or so on the Sagan-Kardashev scale). It took roughly 250 years of industrialization to bring us to the civilization we have today, so we can imagine that it will take about the same amount of time to bring us to the next stage of civilization.

20.3   TYPE I CIVILIZATIONS

A Type I civilization has come to mean one that has complete control over all of the resources available to the planet. Resources is a big concept and could mean a lot of different things; for example, should a Type I civilization be able to control the planet's weather? However, if we just pay attention to energy usage, things become simpler.

The most valuable resource on Earth is sunlight. Don't believe me? Agriculture is impossible without it. Without sunlight, everyone dies; actually, all life on Earth dies, because it is completely dependent on the energy we receive from the Sun. Earth is not a closed system: without an input of energy from the Sun, life can't work. So, in principle, in controlling all of a planet's resources, a Type I civilization will control (harvest, use, whatever) all of the sunlight that intercepts the Earth.

The average insolation for the Earth is about 250 W/m^2 at sea level. The total power of sunlight reaching the Earth is just this number multiplied by the total (projected) area of the Earth, 1.3×10^{14} m^2. So the total power available from sunlight is 250 W × 1.3×10^{14} m^2 = 3.2×10^{16} W. Give or take, this is the energy usage of a Type I civilization on the Sagan-Kardashev scale. But we have to ask whether it is even possible for our world civilization to reach this level of power use, or, if possible, whether it is wise to do so.

First, there are many reasons why we might not be able to harvest all the power the Sun sends us. One obvious reason is that about 70%

of the Earth's surface is water. While all things are possible for those who believe, it's hard for me to think that our descendants will want to plate over the surface of the oceans with solar cells. Solar cells are also not perfectly efficient—the best that exist today run at about 20% conversion efficiency of light into electricity. I suspect also that not more than 10% of the land surface could be effectively covered with solar cells before there would be extreme consequences to Earth's ecology. All three factors would reduce the amount of available power from sunlight by about two orders of magnitude from the value given above, to a level roughly ten times greater than world energy consumption today. Not bad, but far from the godlike beings we want to ascend to.

Of course, there are other power sources available. At the present time, the only plausible alternative power source is controlled thermonuclear fusion. In principle, the world's supply of deuterium could furnish us with unlimited energy, enough to make everyone rich. There is an issue with generating that much power, however: ultimately, any energy generated for running our complex civilization is turned into heat. If we use sunlight for power, it isn't that bad, as the sunlight hitting the Earth heats it anyhow. But generating an enormous amount of power using fusion has the potential to directly heat the Earth by quite a bit. How much? Well, the Earth is maintained at about 300 K simply from heat from the sun. If we generated the same amount of heat using another power source, the temperature of the Earth would increase by a factor of roughly $2^{1/4}$, or to about 360 K (86.5° C or 188° F), much too hot to sustain life. This is disregarding any other pollution such energy generation would create. Solar power doesn't have this problem, mostly: the sunlight absorbed by the Earth turns into heat anyhow, so it doesn't matter if it takes an extra step to generate electricity. (The "mostly" has to do with the fact that if extensive solar cells change the Earth's effective albedo, the Earth's mean temperature will change.)

If you want that much energy, you must pay the price for it. Ultimately, a Type I civilization requires a completely controlled ecology: it would require enormous changes in the Earth's biosphere, to the point that most animals (including humans) would need to be extensively changed in order to survive. Like many science fiction ideas, this concept seems to have originated with Olaf Stapledon in his novels *Last and First Men* and *Star Maker* [225]. In *Star Maker*, the galactic civilizations genetically engineer their citizens into energy-efficient hyperintelligent insect-like beings in order to make the most effective use of the energy given out by dying stars at the end of time. As many authors have suggested, there are other ways this scenario could play out: humans could all "download" themselves into computers so that we could make whatever changes we wanted to the corporeal world without worrying about consequences (a theme pursued in several of Chris Egan's novels, among others); the computers could wipe us out (or try to), as in the *Terminator* movies, and do the same thing (this is explored in detail in the novel *Hyperion*); we could genetically engineer ourselves to live in the extremely changed environments resulting from our advanced technology, as in Bruce Stirling's *Schismatrix* novels; we could accidentally do it to ourselves, and end up with a Type 1 civilization by default (this is more or less the plot of Greg Bear's novel *Blood Music*), or some combination thereof. An excellent article by Hans Joachin Schellnhuber and colleagues addresses these issues in a very readable manner [211].

20.4 MOVING UPWARD

> If we get a good start,
> We can take Mars apart,
> If we just find a big enough wrench...
> —"HOME ON LAGRANGE," TRADITIONAL FILKSONG

So now we control the resources of an entire planet. What do we do now? We ascend higher, that's what!

Human colonies in space or on other planets is the next logical step. We've discussed such colonies in other chapters, although of course one can imagine creating such colonies even before the planet reaches a Type I civilization. However, there is a next step: the terraforming of a world. The idea is to take a (presumably) lifeless planet such as Mars and make it habitable for people. What "habitable" implies exactly is a little loose, but generally speaking, the idea is that a human could live out in the open on such a world. Of course, one can imagine modifying humans through genetic engineering or creating a cyborg or some such so that they could survive there, instead of modifying the entire planet. Or perhaps some sort of combined approach, in which we changed both the people and the planet, would work.

It's hard to imagine what such large-scale changes would mean, or if such a civilization could control them. We are currently conducting an uncontrolled experiment in which we are increasing the average temperature of our planet by increasing the atmospheric retention of heat from the Sun, and it has already had a measurable effect on our ecology. Paul J. Crutzen and Eugene F. Stoermer coined the term "anthropocene" to describe the current epoch, which humans entered when the Industrial Revolution began to change the world's ecology and weather patterns [61]. Some even extend the term back to the beginning of the invention of farming, about 20,000 years ago.

It seems clear that the more power a civilization uses, the more it will change the environment around it, and probably in many unforeseen ways. The global weather system is a classic example of a chaotic system—small changes to the inputs driving the weather can have big changes in its outputs. This has been known on some level since 1963, when Edward N. Lorenz put together a simple model of convection in Earth's atmosphere [154]. To quote Lorenz, "When our results... are applied to the atmosphere . . . they indicate that the prediction of the sufficiently distant future is impossible by any method." This is occasionally called the *butterfly effect*: the flapping of a butterfly's wings in China can cause a hurricane to sweep across Texas. The Earth's ecology is another example of a chaotic system, if one with a much longer response time: most mathematical models of small parts of Earth's ecology, such as predator-prey models, involve strongly coupled nonlinear differential equations, the very hallmark of chaos. To make matters worse, the ecology is complex enough that it isn't clear whether most of the current mathematical simulations of it are close to reality [231].

20.5 TYPE II CIVILIZATIONS

On the Kardashev scale, Type II civilizations control the resources of an entire solar system, or, using the formula given above, have access to power of order 10^{26} W, the power output of an "average" star like the Sun. Getting from Type I to Type II may present something of a problem. I'm not sure I know of any science fiction novels that have directly addressed this issue. The main problem is this: harvesting all the power from the Sun involves dismantling planets. The issue is creating a screen to capture the light from the star.

Larry Niven's novel *Ringworld* is an interesting example of this approach. The Ringworld, whose dynamics are discussed in chapter 19, is an enormous ring around a star with radius just about equal to Earth's mean distance from our Sun. It's spun at an enormous rate, and the scale of the thing implies that a Type II civilization built it. The Ringworld civilization controls the power of its star; there is a "shadow-square" system that not only provides day and night to the Ringworld, it also defends it against outside threats using controlled solar flares. In an odd way, Niven's story provides an example of what I mentioned above: people might change in unexpected ways in order to survive in such an engineered environment. On the Ringworld, the fall of civilization leads to humans evolving into a variety of unoccupied ecological niches, including ghouls, a carrion-eating subspecies, who fill the roles of hyenas on Earth [183].

Civilizations somewhere between Type II and Type III are common in the media and in older science fiction, even though not called by that name. Any Federation of Planets or Galactic Empire, so beloved of older writers, is by definition higher than Type II. Don't believe me? Let's consider a small point from the television show *Star Trek: The Next Generation* and its sequel series: the replicator.

The replicator was presumably an offshoot of transporter technology. We've discussed teleportation in an earlier chapter. I tend to think it impossible, for various reasons, but what the heck: let's assume it works somehow. As all fans of the series know, Captain Picard tells a little box sitting in his office, "Tea, Earl Gray, hot," and moments later a cup materializes in it. (And contains, one is tempted to say, "a liquid that was almost, but not quite, entirely unlike tea." Thank you, Douglas Adams! [19]) Whether the cup is prepared elsewhere and transported to the office or created whole is never said. Presumably the latter, as the Holodeck seems to work on the same principles. Now, a cup of tea has a mass of about 250 grams. This represents an energy equivalent of

2.25×10^{16} J. This is about the same amount of energy currently used by *the entire United States in two and a half hours.* If there are seven billion people in the Federation of Planets, each wanting a cup of tea, then this is about the same amount of energy that the Sun radiates in half a second! Of course, this is probably a low population estimate. In any event, the tiny little replicator indicates that the Federation is able to handle enormous amounts of energy. If we assume that all the food in the Federation is generated in this manner, then the amount of power required approaches the total output power of several stars.

This seems like a damn-fool way to make a cup of tea. No technology is 100% efficient; the waste heat generated from using a replicator would probably melt (or vaporize) any container one put it in. Most science fiction TV shows and movies don't deal with these issues honestly, in my opinion. The movie *Bladerunner*, loosely based on Philip Dick's *Do Androids Dream of Electric Sheep?*, features a 1940s noir world on the brink of collapse. However, there are advertisements in the film for volunteer colonists to other worlds, which implies a very energy-rich civilization. The two aren't mutually exclusive, but they are difficult to reconcile.

20.6   TYPE III CIVILIZATIONS

Stanislaw Lem wrote a satirical story about aliens who had reached the HPLD, "highest possible level of development." They basically sat around doing nothing, creating such miracles as square planets, and the like. They did nothing because, as they explained to one explorer, anything they did would represent a step down. I'm not sure it's possible to write meaningful stories involving Type III civilizations for that reason: once you get to a certain point, Clarke's dictum about sufficiently advanced civilization seems a bit tame. "Any *really* advanced technology is God-like" might be closer to the mark. In this book I have tried to stay close to what we know today and can extrapolate based on currently understood laws of physics. Any Type III Kardashev civilization, one that can effectively exploit the resources of an entire galaxy, must be long-lived on a scale that is unknown in human history. If we believe that the speed of light is the ultimate barrier, then it will take millions of years to spread across the galaxy. Even if we believe that faster-than-light travel is possible, this is still probably true. It also must be farsighted in a manner in which no contemporary human civilization is, and have clearer goals. *This is because the only justification for such a society is the ultimate extreme long-term survival of that society.* Nothing else makes sense. In earlier chapters we saw that any form of space travel is infinitely easier with unmanned probes: putting people on other planets is expensive to the point of insanity, and horrifically dangerous. However, it is the only means by which the human race can survive long term, meaning beyond a few million years. That is the subject of the next chapter.

CHAPTER TWENTY-ONE

A GOOGOL YEARS

21.1 THE FUTURE OF THE FUTURE

In this final chapter of the book I would like to examine the grandest theme in science fiction, the future of intelligent life in the universe. And by future I mean the *far* future. In previous chapters we've examined what truly advanced civilizations are capable of, and how they can potentially modify their environments in extreme ways. In this chapter I would like to examine how long such extraordinarily advanced societies can hope to last. It will dwarf all the history of the universe to this point by a long shot. Let us consider the long-term survival of humanity and intelligent life in the cosmos on the timescale of hundreds of millions of years to billions of years to even longer.

21.2 THE "SHORT TERM": UP TO 500 MILLION YEARS OR SO

There is no guarantee that humanity will survive the next hundred years, let alone the next hundred million. Even if we don't do ourselves in, natural climatological cycles may do the trick for us. Because of slow, periodic changes in the Earth's orbit, Earth's climate goes through cycles of glaciation followed by thawing over the course of about 100,000 years: about 80,000 years of glaciation, followed by about 20,000 years of interglacial periods. All of current human civilization has been encompassed by the last interglacial period, as stable human society is possible because of the invention of agriculture. This was only possible after the glaciers retreated some 15,000 years ago. The biggest short-term threat to humanity is global warming. If we beat that, however, our descendants several thousand years from now will have to face global cooling. It isn't entirely clear that humanity could survive an ice age, although there are a lot of imponderables in that statement. There are no fundamental physical reasons why we couldn't, however, so I'm going to take an optimistic view, and expand the timescale by a factor of 1,000. The problems happening over this time period are easier to discuss from a physics standpoint.

Over the next few hundred million years, the biggest threat to humanity is expected to come from comet or asteroid impacts like the ones that led to the demise of the dinosaurs. The dinosaurs were killed off 65 million years ago when a comet or asteroid roughly 20 km in diameter hit the Earth. With a speed of roughly 40 km/s at impact and a mass of about 10^{15} kg, the kinetic energy of the event would have been about 10^{24} J. This is billions of times larger than all of the energy that would be liberated if all the world's nuclear arsenals were exploded at once. These impacts have been the subject of fiction: *Lucifer's Hammer* and the movies *Deep Impact* and *Armageddon* deal with impacts like this. In an earlier chapter we looked at the damage a nuclear war would do. The biggest problems would come from the nuclear winter caused by particulates blocking sunlight. An impact like this one would block sunlight for years, causing a massive dying off of most life on Earth. One theory of the history of mass extinctions in Earth's paleontological history posits impacts like the dinosaur-killer happening roughly every 100 million years.

Even smaller impacts, which occur correspondingly more frequently, could destroy civilization and possibly all life on Earth. Astronomers have seen an impact like this one in 1993, when comet Shoemaker–Levy 9 hit Jupiter. This comet, which calved into several separate sections before impact, was smaller than the dinosaur-killer, but its impact energy was still larger than the world's total nuclear arsenal.

It seems likely that within the next few hundred million years Earth will suffer a similar impact. Large impactors still exist in the Solar System. The orbits of a large number of asteroids pass near Earth. There was a scare about a decade ago when the asteroid 99942 Apophis was predicted to have a non-zero chance of hitting Earth in the year 2031.

21.3 THE "MEDIUM TERM": UP TO ABOUT 10^{13} YEARS

Perhaps a sufficiently advanced civilization will want to move the planet. If so, it can turn asteroid and comet impacts into a positive good. There are good reasons to want to move the Earth: in one billion years the luminosity of our sun will increase by 10% [130]. This will increase Earth's temperature to the point that life will not be possible. Our descendants (or our replacements) might want to move Earth farther away from the Sun to keep it cool. At that stage, they would only need to move it outward by about .05 AU to keep the flux of light from the Sun about the same as it is now. Much later, five billion years or so from now, the Sun will exhaust the hydrogen in its core and swell into a red giant. The luminosity will then be many thousands of times what it is now, and the sun itself will swell until it is larger than the current orbit of Mercury. So, how do we move a planet?

Interestingly enough, the energy it would take to move Earth farther from the Sun by 1 AU is comparable to the energy needed to dismantle it. If we want to move a planet of mass M_p from a circular orbit of radius r_1 to radius r_2, the change in the total energy of the planet-Sun system is given by

$$\Delta E = \frac{1}{2} \left(G M_s M_p \right) \left(\frac{1}{r_1} - \frac{1}{r_2} \right). \tag{21.1}$$

The factor of 1/2 comes from what is called the "virial theorem." Moving Earth from 1 AU to a distance of 2 AU has $\Delta E = 2 \times 10^{33}$ J, equivalent to the Sun's total energy output for two months. A few authors, including Freeman Dyson, have given some thought to how this might be done.

One interesting paper on the subject uses a method similar to the gravitational slingshot method [142]. A planetoid of mass m falling from a large distance away from the Sun to a distance r away from it gains kinetic energy equal to

$$K = \frac{G M_s m}{r}. \tag{21.2}$$

Let $r = 1.5 \times 10^{11}$ m $\times a$. That is, let's express the distance in astronomical units. Then

$$K/m = 8.85 \times 10^8 \text{ J/kg} \times \frac{1}{a}.$$

The velocity of the planetoid as it reaches a distance a from the Sun is

$$v = 42 \text{ km/s} \times \frac{1}{\sqrt{a}}. \tag{21.3}$$

In principle, the impact or near collision of such a planetoid with Earth would change its velocity by an amount of order

$$\Delta v \approx \frac{m}{M_E} v \approx 42 \text{ km/s} \times \mu, \tag{21.4}$$

where $\mu = m/M_E$. This is the same principle by which a spacecraft can increase its velocity in a close orbit around a planet. However, in this case the situation is reversed: we want to change the planet's motion, not the spacecraft's. In the original case, the speed of the spacecraft can be changed dramatically because the planet outweighs it by a factor of more than 10^{22}. In this case, the change in the Earth's orbital speed will be tiny, because even a very large planetoid will only have a small fraction of Earth's mass.

There is a lot of icy debris orbiting the Sun at distances from about 30 AU out to nearly 100 thousand AU. The Kuiper Belt, extending from about 30 AU to about 200 AU, is the home of the dwarf planets Pluto and Eris, both of which have masses of order $2 \times 10^{-3} \times M_E$. There are probably trillions of smaller pieces of ice and dust and rock; the total mass of the belt is estimated to be about 30 M_E. One can imagine a sufficiently advanced civilization arresting most of the orbital motion of pieces of this debris and letting it "fall" into the inner Solar System. If one carefully chose its orbit so that it passed close to the Earth, it would be possible to increase Earth's orbital speed, bringing it farther out from the Sun. It would take quite some time to do this. An object falling from a distance of 100 AU from the sun would take about 175 years to reach Earth's orbit. The authors of the paper estimated that one would need about 10^6 such near collisions to move Earth's orbit outward to 1.5 AU. This assumes an average value of $\mu = 1.7 \times 10^{-6}$, or $m = 10^{19}$ kg. The required net Δv change is of order 10 km/s. One issue mentioned in the paper is that the timing of the orbital approach of these planetoids as they passed Earth would need to be down to the minute. Again, each maneuver would take centuries, or even millennia [142].

The net energy needed for this maneuver is of order 10^{33} J. We have about one billion years in which to move the Earth. This means the average rate at which we need to expend this energy is

$$P = \frac{10^{33} \text{ J}}{\left(10^9 \text{ yr} \times 3.16 \times 10^7 \text{ s/yr}\right)} \approx 3 \times 10^{16} \text{ W}.$$

This is three orders of magnitude above the net energy expenditure of our current civilization, or an energy expenditure rate of a Kardashev Type I civilization. This makes sense: one cannot conceive of moving a planet without access to vast reserves of energy.

This maneuver was meant to protect the Earth's climate from the Sun's luminosity changes as the Sun evolved along the main sequence. After it leaves the main sequence, things get trickier. As the Sun turns into a red giant star, its luminosity will temporarily increase to over 1,000 times what it is currently. However, its luminosity will then decrease after the "helium flash" to about 100 [130, pp. 468–470]. At the high value of the luminosity, the Earth would need to be moved to 30 AU from the Sun (roughly the orbital distance of Pluto). After that, it would need to be moved back to 10 AU. This would take place on a timescale of hundreds of thousands or millions of years, not billions.

This means that the power expenditure rates would need to be thousands of times higher than what we just considered. To a very long-lived observer, Earth would appear to be in some cosmic Ping-Pong game. The power expenditure rates would require a Kardashev Type II civilization, although at that point it might just be easier to move everyone to a new star system.

21.4 THE "LONG TERM": UP TO A GOOGOL YEARS

And AC said, "LET THERE BE LIGHT!"
—ISAAC ASIMOV, "THE LAST QUESTION"

I want to make my underlying assumptions here clear: to discuss the far future of the universe, one must have a model for the evolution of said universe in mind. The model I am considering is the best one astronomers currently have. It is referred to as the "inflationary Big Bang consensus model." The main ideas of this model are:

1. The universe began in a big bang some 13.7 billion years ago. We do not know if there was anything before this (if that word has meaning) or if there are other universes like our own.
2. Shortly after the universe began, it went through period in which it expanded rapidly from the size of a proton to the size of a basketball. This is known as "inflation." Most of the details of inflation aren't well understood, but there seems to be very good evidence that it happened.
3. The mass-energy content of the universe is distributed as follows: about 3% of the matter in the universe is ordinary matter, such as we are familiar with on Earth. About 90% of that is hydrogen. Of the other 97%, about 23% is "dark matter"; scientists don't know what it is, except that it apparently doesn't act much like regular matter (i.e., it doesn't interact strongly with other matter except via gravitational interactions). The other 74% is "dark energy"; we *really* don't know what that is.

4. The universe is "flat," in the sense that the total amount of matter/energy in the universe (apart from gravitational self-attraction) exactly balances out the gravitational self-energy of the universe.

5. However, even though the universe is balanced so precisely, the dark energy is accelerating the expansion of the universe so that instead of slowing down owing to gravity as it expands out, it is accelerating at an ever-increasing rate.

This consensus model has been developed since 1998 when observations of far-away supernovas led to the measurement of the dark energy and the acceleration of the universe. This isn't the place to go into the evidence for the consensus model. Suffice it to say that there is a lot of it. If you are interested, I would suggest either a basic astronomy textbook such as *21st Century Astronomy* or a popular book such as Mario Livio's *The Accelerating Universe* [130][153]. Make sure that any astronomy textbook you read was published after 1998! The cosmology section of any textbook published before then is completely out of date.

The implication of the accelerating universe is that it will never go through a "Big Crunch" in which the universe halts its expansion and recollapses. (I preferred Douglas Adams's term "Gnab Gib" over "Big Crunch," but the point is moot now.) These new discoveries invalidate narratives like *Tau Zero* that are predicated on a cycle of Big Bang–Big Crunch–Big Bang, and so on. Such a model was aesthetically pleasing, but it seems that the universe is just not like that.

Among science fiction writers, Olaf Stapledon was probably the first to consider the very long-term future history of the universe. We've discussed *Star Maker* in an earlier chapter [225]. Stapledon implicitly used the idea of the open universe in *Star Maker*; at the end of time, the universe consisted mostly of galaxies too far away to communicate with each other, each consisting of a few remnants of the once great stellar populations. The long-term evolution of the universe has made its way into a number of science fiction books since Stapledon. Most writers have seemed to favor the "closed universe," or cyclical universe, as *Tau Zero* does, as it seems to envision an infinite future. Even if humanity can't survive forever, perhaps some form of intelligent life can evolve in the next cycle of the universe.

Main-sequence lifetime is tied to stellar mass. For example, in about four or five billion years our Sun will move off the main sequence and swell into a red giant whose luminosity will exceed 100 times its present value. As it burns through its nuclear fuel, it will cool and contract to a white dwarf whose total luminosity will be initially about 10^{-3} what it is today, and will further cool over time. Higher mass stars than the Sun burn through their nuclear fuel quickly and end their lives in spectacular supernova explosions. Low-mass stars live significantly longer than high-mass ones. Luckily, most stars in the sky are low-mass M-class stars with low luminosity but very long lives. This allows the possibility (as in Asimov's story) of moving to a new star once the Sun has evolved into a white dwarf. Stellar lifetimes are given by the approximate formula

$$\tau = 10^{10} \text{ years} \times \frac{M}{L}, \tag{21.5}$$

where M is the stellar mass and L its luminosity. From chapter 14, an M7 star has $M = 0.08$ and $L = 0.0025$, leading to a main-sequence lifetime of 3.2×10^{11} (320 billion) years. The estimate is probably an underestimate of their true lifetime. Our Sun will run through only about 10% of its hydrogen before swelling into a red giant. This is because the Sun's core isn't convective, that is, it isn't stirred around by currents created by temperature gradients. This means that the core, where fusion takes place, can't get any new material to fuse once the core's supply is gone. M-class red dwarf stars have fully convective cores, meaning that new fuel is mixed in constantly. This increases their main-sequence lifetime above our crude estimate [47]. Stars at the lowest end of the mass range may have lifetimes near 10^{13} years.

The long-term issue we face is similar to the short-term issue our civilization is facing today: resource depletion. We are rapidly using up the Earth's reserve of fossil fuels now. In this imaginably distant future we will be using up the resources of the stars. If human civilization lasts for a few billion years, I feel confident in predicting that we will have some form of interstellar travel by then, because we will need it. Maybe we can move our entire planet to a new solar system. Both Freeman Dyson and Larry Niven have speculated on methods of doing this.

21.5 BLACK HOLE–POWERED CIVILIZATIONS

When all the stars die, that's the end, right? All other sources of energy are gone, aren't they? Well, maybe not. Black holes will be around for a long time after all the stars grow cold, and they offer the potential of providing energy in those end times (which may last for a much, much longer time than all of the other eras in the universe).

A black hole may seem like a bad source of energy. After all, the common conception of them is that they swallow everything that enters them, and nothing can come out. However, that's not quite true: as we saw in an earlier chapter, black holes radiate away energy (Hawking radiation), even though it is too low to detect directly.

There are two ways to retrieve energy from things dropped into black holes:

1. By electromagnetic radiation from the things dropped into them via heating in the accretion disk; and
2. By gravitational radiation.

The first process is relatively straightforward: the radius for the event horizon of a (nonrotating) black hole is determined only by the mass of the black hole:

$$R_{BH} = \frac{2GM}{c^2} = 3\,\text{km} \times \frac{M}{M_s},\tag{21.6}$$

where M_s is the Sun's mass. Therefore, a small black hole ten times the mass of the Sun would have an event horizon 15 km in radius. Imagine dropping a 1 kg object from a long distance away toward the black hole and abruptly stopping it three radii away. In its fall it would acquire kinetic energy approximately equal to $1/6\ mc^2 = 1.5 \times 10^{16}$ J. I am using the Newtonian formula here, which is an approximation to the full relativistic formula. I chose three radii from the black hole for two reasons:

1. Three radii away is the closest distance at which stable orbits are possible [235].

2. It is also the closest radius at which I feel comfortable using Newtonian formulas to estimate the energy liberated.

This is an enormous amount of energy; we could power all of current-day America's energy needs using 600 kg of trash, assuming we could reclaim the energy liberated at 100% efficiency. This is essentially the method that astronomers have used to find black holes that have evolved from large stars: if a black hole is in close orbit with a normal star, gases from the star will be funneled into the black hole. Frictional heating of the gases as they fall in leads to them reaching temperatures in the millions to hundreds of millions of degrees, radiating away enormous amounts of energy in the x-ray spectrum.

A more subtle way of generating energy using black holes is via gravitational waves. Gravitational waves are literally ripples in the fabric of space and time. It has been predicted that merging black holes are a strong source of gravitational waves. This is one of the key things the Laser Interferometer Gravitational-wave Observatory (LIGO) gravity wave detector is looking for. In principle, up to 50% of the mass-energy content of any junk dropped into a black hole can be recovered as useful energy [236]. This is a much higher efficiency than any other known source of energy.

The science fiction idea of powering objects using black holes dates back to the 1980s, if not earlier. The earliest use I know of is in the McAndrew chronicles, a series of stories written by Charles Sheffield centering on the eponymous astrophysicist [217]. In these stories, mini-black holes are used to power spacecraft, and presumably other things as well. The discovery of Hawking radiation makes the stories obsolete, as the mini-black holes of the stories would evaporate too quickly to use. The final episode of the new *Battlestar Galactica* series had the Cylons in orbit around a black hole, powering their civilization by throwing trash into it.

This is the method by which a civilization past the death of all of the stars could get energy: take your unused trash and toss it into the black hole. This leads to various baroque speculations and plot ideas. One can imagine some far-distant civilization powering their energy needs for the upcoming year by tossing a sacrificial virgin or two into the black hole. It would certainly work better than tossing them into a volcano.

21.6   PROTONS DECAY—OR DO THEY?

One thing that may cut short our joyous spree into the forever is the possibility that ordinary matter may softly and silently vanish away. This Boojum is the Snark of proton decay.

The proton is one of the three building blocks of ordinary matter. Of the other two constituents, the electron is the lightest—about 1,800 times lighter than the proton. It is a stable particle. The neutron is not: the neutron is a charge-neutral particle made up of one "up" quark, with charge $+2/3$ of the electron charge, and two "down" quarks, each with charge $-1/3$. When outside the nucleus, the neutron can decay into a proton, electron, and antielectron neutrino; the decay converts one of the down quarks into an up quark, which is what turns the neutron into a proton. The proton is two ups and one down, with net charge of $+1$. The neutron decay time is long by physics standards, taking tens of seconds. Because protons aren't elementary particles either, the possibility exists that they could decay into lighter particles as well.

This has never been seen experimentally, but some theories of physics such as string theory predict it. If protons decay, they take a very long time to do so. The universe has been around for 13.7 billion years, so this puts a lower limit on the time it takes. If protons decayed much faster than this, we wouldn't be around. Because it's so hard to calculate anything using string theory, there are no very good predictions for the proton decay rate. However, experiments set the proton lifetime as being greater than 10^{34} years. A friend of mine once made the comment that *that* was a pretty good definition of forever, but it's not good enough for our purposes!

21.7   A GOOGOL YEARS—ALL THE BLACK HOLES EVAPORATE

If protons don't decay, then the ultimate lifetime of life in the universe may be set by the timescale it takes for black holes to evaporate. Black hole evaporation is a quantum mechanical phenomenon. If we attempt to put a particle inside a black hole, quantum mechanics tells us there is a small probability that one will find it outside the hole. This is because of the Heisenberg uncertainty principle. We work out the basics of this phenomenon in the web problems. The bottom line is that in 1974, Stephen Hawking showed that black holes aren't completely black. They act as *blackbody radiators*, though at very low temperature. A small amount of energy leaks out, and one can even assign a temperature to them:

$$T_{BH} = \frac{hc^3}{16\pi^2 G k_B M} = 6.2 \times 10^{-8} \times \left(\frac{M_s}{M}\right) \text{K}, \tag{21.7}$$

where M is the mass of the black hole, and the other terms have been defined previously in this book. A black hole with the mass of the Sun would have a temperature of only 6×10^{-8} K, and larger ones would have lower temperatures. Still, the standard formulas for blackbody radiation apply, even to such exotic objects. One can show that the rate at which the black hole radiates away energy is proportional to $1/M^2$. A completely isolated black hole will spontaneously radiate away energy. Its mass will decrease. As its mass decreases, it will radiate away energy at a higher rate, which will cause it to decrease in size more rapidly, leading to an explosion of energy in the last few seconds of its existence. One can calculate the time the black hole will last from this formula:

$$\tau_{BH} = \frac{32,768\pi^2}{3} \frac{G^2}{hc^4} M^3 = 2.2 \times 10^{67} \times \left(\frac{M}{M_s}\right)^3 \text{yr}. \tag{21.8}$$

A black hole with the mass of the Sun will far outlast all of the stars. A more typical black hole with a mass ten times that of the Sun will last for 2×10^{70} years. But this is peanuts compared to the largest black holes around.

At the center of each galaxy are ultramassive black holes whose mass can range from about a million times to several billion times the mass of our Sun. Our own galaxy has a relatively modest one with a mass of only 30 million Suns. A civilization in an artificial planet or Dyson net in orbit around this could potentially last for more than 10^{89} years. This dwarfs the current age of the universe by a huge margin, but we can do better.

The largest known black holes have a mass of more than 10^{10} solar masses. They are billions of light-years away, of course, but we have all the time we need to get there. A 70 billion solar mass black hole would have a lifetime of 10^{100} years—a googol years. There are currently no known black holes with this mass. The largest, discovered in April 2011, has a mass of 21 billion solar masses. However, I'm going to assume we can find a larger one, because writing a googol years is cooler than writing 10^{98} years.

To indicate how long a time this is, the current age of the universe is about 10^{10} years. If the age of the universe so far was represented by, say, the mass of a proton, a googol years would be represented by ... what? The mass of all the grains of sand on all the world's beaches? No. The mass of the Earth? No. The Sun? *No.* The mass of our universe? *NO.* A googol years would be represented by all of the visible mass in *ten billion universes just like the one we are in right now.* By the way, one great satisfaction in writing this section is that I have finally found a practical use for the term googol, which has not had much application in mathematics or physics to date, despite a fair amount of commercial (if badly spelled) success.

I am not the first person to have made these speculations. One of the interesting things about writing this book is that certain names keep popping up. Larry Niven and Poul Anderson are the two science fiction writers whom I have turned to for inspiration many times; among scientists, Freeman Dyson is the clear standout. In a 1974 paper he did the same thing that I am doing here: he calculated how long intelligent life in the universe could last [73]. His conclusions were more optimistic than mine: he concluded that life in the universe could last indefinitely by going through cycles of hibernation and activity, using less and less energy on each active cycle. One point: because of the paper's date, a number of ideas in it don't reflect current ideas in cosmology. In particular, the consensus model invalidates a number of his ideas. The paper "A Dying Universe: The Long-Term Fate of Astrophysical Objects" by Fred C. Adams and Gregory Laughlin is a more up-to-date analysis of this idea; I recommend it for the science fiction writer interested in *very* far-out ideas, as it contains a trove of data and formulas on the subject of the eventual fate of the universe [20]. The article is slightly out of date, as it predates the supernova measurements leading to the concept of the accelerating universe. However, its last section discusses the fate of the universe with non-zero cosmological constant.

21.8   OUR LAST BOW

> My pen halts, though I do not. Reader, you will walk no more with me. It is time we take up our lives.
>
> —GENE WOLFE, *THE CITADEL OF THE AUTARCH*

Jack Vance, Isaac Asimov, Gene Wolfe, Neil Gaiman, and many, many others have stories set at the end of time. This usually means at the end of the Earth's lifetime, but some have gone much, much farther than that. As I said at the outset, this book is not meant to be predictive. The same can be said for science fiction itself. I do not expect humankind to last for a googol years; even if it does, it would not exist in any form recognizably human for even a tiny fraction of that time. However, even in the most far-flung stories, humans must remain human if we (as humans) are to sympathize with (or even understand) their actions. This is perhaps the greatest limitation of the literature: the hopes and dreams of one little species don't amount to a hill of beans in this crazy universe. Or as Neil Gaiman put it, you can have happy endings as long as you end the story early enough.

I'm finishing this book with a quotation that speaks to me in a deep way. Let it be a metaphor for the best in science fiction as well as the best in humanity. It is from the great French mathematician Henri Poincaré:

> Geologic history shows us that life is only a short episode between two eternities of death, and that even in this episode, conscious thought has lasted and will only last a moment. Thought is only a gleam in the midst of a long night.

> But it is this gleam which is everything [192].

ACKNOWLEDGMENTS

Any book is the work of many people, not just the author. Much of the credit for what you are reading goes to a lot of other people. Any mistakes in it are of course mine alone.

First and foremost, I would like to thank my editor at Princeton University Press, Vickie Kearn, whose support and enthusiasm made this project possible. I would also like to thank Quinn Fusting, Natalie Baan, and Marjorie Pannell for their invaluable help, and all of the other staff at the press and elsewhere who worked on the book.

I would like to personally thank one of my favorite science fiction writers, Larry Niven, for letting me quote from his letter to Roger Zelazny. I found the quotation in Zelazny's papers, held in the Azriel Rosenfeld Science Fiction Collection at the Albin O. Kuhn Library at the University of Maryland–Baltimore. I thank the staff of the library, especially the curator, Thomas Beck, for their help. Much of this book was researched and written at the Library of Congress in Washington, D.C.; I thank the staff of the main reading room for their help in retrieving books and articles, and for maintaining such a wonderful place to work.

The science fiction writers who have influenced me are too numerous to be mentioned here by name in every case, but two must be singled out for special credit: Poul Anderson, to whom the book is dedicated, for his essays on writing science fiction, and Olaf Stapledon, who provides the best example of what speculative fiction can do. Among scientists,

Freeman Dyson has probably seeded more ideas used in hard science fiction today than any other living physicist.

My colleague, Mark Vagins of the University of Tokyo, had the initial idea for the section "Thrown for a Loop" in chapter 7. Mark is one of the smartest people I know, and has been a close friend since high school. It is with his kind permission that the analysis is reprinted here. Anthony Bowdoin "Bow" van Riper is an expert on science and popular culture. He gave me invaluable information on infrastructure costs associated with the Space Shuttle program. Raymond Lee, a physicist and meteorologist at the Naval Academy, measured the luminous flux of several candles for chapter 3, "Why Hogwarts Is So dark." He also discussed subtle points concerning different types of photometric measurements and gave me advice on calculating luminosity and luminous flux from various light sources. I also thank Zeke Kisling, who gave me invaluable advice on formatting and presentation in the book.

The anonymous reviewers for Princeton University Press pointed out several issues with the book that needed correction. I thank them for their diligent reading, and their support for the book. I also wish to thank Larry Weinstein and Paul Nahin for their reviews of the original proposal.

I tested out a lot of the ideas in this book in the class "The Science of Science Fiction" offered at St. Mary's College in the spring semester of 2009. I thank the students in the class, Roger Ding, Adam Hammett, Galen Hench, Devon Jerrard, Malory Knott, James Moderski, David Panks, Abby Taylor, and David Tondorf-Dick, for beta testing much of the material in the second and fourth sections of the book. I'm sorry there wasn't enough room in the book to include their great class projects. I also ran a much earlier version of this class at Cleveland State University in 1996. I thank the students who took that course, although I no longer have their names.

My mother, Louise Adler, read one of the earlier drafts of the book and gave me a lot of support and several practical suggestions about organizing the work. Finally, and most important, I thank my wife, Karen, and our daughters, Alexandra and Cassandra, for their love and support at all times, especially while I was writing the book. *Scio quid sit amor.*

APPENDIX: NEWTON'S THREE LAWS OF MOTION

I've chosen to concentrate on known laws of science in this book. This is unlike many books on science fiction, which concentrate on the unknown, or on more speculative laws. Because I'm a physicist, most of the book is devoted to the physics in science fiction. However, to understand this, it is essential to know the basics. In this case, that means understanding the laws of motion as set forth by Sir Isaac Newton 300 years ago. This appendix offers a very short introduction to the laws of motion and to two special cases of accelerated motion: constant acceleration and motion around a circle at constant speed. I am assuming that my readers have some concepts of what force, velocity, and mass are, even if they are not exactly what physicists mean by those terms.

There are three important concepts that need to be discussed before we can introduce the laws: displacement, velocity, and acceleration. *Displacement* of something means moving it from one place to another. It includes both the distance moved and the direction it moved in, specified in relation to some coordinate system. *Velocity* is how fast an object is moving, plus some indication of the direction in which the object is moving. It is the time derivative of displacement:

$$v = dr/dt \tag{A1}$$

where r is displacement and v is velocity. Boldface indicates that we need to include the direction the object is moving in as well as the distance it moved when specifying both velocity and displacement. There are a number of ways to do this. Readers who are interested in a deeper understanding should consult an introductory physics textbook such as *Fundamentals of Physics* [246].

Acceleration is the rate at which velocity changes. From calculus, we specify this as:

$$a = dv/dt, \tag{A2}$$

where a is acceleration. Because the definition includes direction, if an object changes either its direction of motion or its speed, it is undergoing acceleration.

A.1 NEWTON'S LAWS OF MOTION

Newton's first law: No experiment can be used to distinguish between moving at constant velocity or standing still. This is not the normal way in which Newton's first law is usually written, but it is the most accurate. All motions are relative motions. Because forces produce accelerations, we can only measure changes in velocity, not absolute velocities themselves.

Before discussing Newton's second law, we need to define force. A force is a push or a pull.

Newton's second law: Forces produce acceleration. In mathematical terms,

$$F = ma. \tag{A3}$$

Here, m is the mass of the object. Pushes or pulls produce changes in the speed or direction of an object but don't produce velocity directly. This is one of the subtlest aspects of Newton's second law.

Newton's third law: If object A exerts a force on object B, B exerts an equal but oppositely directed force on A. If I push on the wall with a force of 100 N (about equivalent to a weight of 20 pounds), the wall pushes back on me with an equal force in the opposite direction. The third law leads to the conservation of momentum discussed in chapters 2, 6, and 15.

Special cases: There are only two cases of accelerated motion considered at any length in this book—constant acceleration and circular motion at constant speed.

> *Constant acceleration:* If a force of size F acts on an object and the force is constant in time and always pointing in the same direction, the object will have a constant acceleration:
>
> $$a = F/m. \tag{A4}$$
>
> I am using plain text to specify the size of the force and acceleration without reference to direction. The object will move in a straight line. If the object starts with zero speed, it will have a speed over time given by the formula:
>
> $$v = at, \tag{A5}$$
>
> and displacement
>
> $$x = 1/2 \, at^2. \tag{A6}$$
>
> It's traditional to use x or y to indicate displacement when the object moves in a straight line. The symbol x is usually used to indicate horizontal motion and y vertical motion such as free fall under the influence of gravity.
>
> *Circular motion at constant speed:* Motion in a circle is accelerated even though the speed is constant because the direction is continually changing. The force needed is directed toward the center of the circle. This type of motion is often called centripetal ("center-pointing") motion because of this. The magnitude of the acceleration is given by
>
> $$a = v^2/R, \tag{A7}$$
>
> where v is the speed and R the radius of the circle. The true force producing the motion (the centripetal force) is directed inward, but someone traveling on the object in motion will seem to experience an outward-pointing centrifugal force of the same size. This centrifugal force is an illusion due to Newton's first law, but it is often useful.

A.2 FUNDAMENTAL FORCES

There are only four fundamental forces, and we experience only two of them (gravitational and electromagnetic) in ordinary life. Equations describing them are used as needed in the text. The forces are the following:

> *Gravity:* We are all most familiar with this force. Surprisingly, it is the weakest of the four fundamental forces by a very large factor. It acts between two bodies which have mass and is always attractive (that is, always pulls objects together).
>
> *Electromagnetic force:* This force exists between charged objects. There are two types of charges, called + (the sign of the proton charge) and − (the sign of the electron charge.) Like charges repel, unlike attract. There are two odd things about this force: first, charges are quantized. Measurable charges come in integer multiples of the electron charge. No one really knows why. Second, the electromagnetic force is stronger than the gravitational force by a factor of 10^{40}. Again, there's no really good explanation why the strengths of the two forces are so unbalanced.
>
> If charges are in motion, they create a magnetic field. If they oscillate back and forth, they create light. These are part of the electromagnetic force as well, but their properties aren't as simple as the force between nonmoving charges. I discuss some of these things in chapter 3, but if you want a more complete explanation, see any first-year physics textbook.
>
> *Strong force:* The nucleus of the atom is composed of protons and neutrons. Therefore, it is composed of a lot of positive charges pushing against each other. The nucleus doesn't explode: some force must hold it together. This is the strong force. It is a force exerted between nucleons; it doesn't extend to very far distances as the electromagnetic or gravitational forces do. It essentially dies off at distances much larger than about 10^{-15} m. The force is really between the quarks inside each of the nucleons. The force between neutrons and protons is just that which "leaks out."
>
> *Weak force:* This is a force responsible for certain types of decays such as the neutron into the proton, electron, and electron-antineutrino.

It is very interesting: unlike the other forces, it can distinguish between right and left, and between antimatter and matter. It is very short range, but may be responsible for the existence of matter in the universe. Because of the asymmetry between matter and antimatter, during the Big Bang a very small (about one in a billion) preponderance of matter over antimatter arose. Most of the matter and all of the antimatter annihilated; the small remnant left over is the matter we see in the universe today.

Derived forces: All other forces are derived from the fundamental ones. Chemical bonds are electromagnetic in origin. They are due to the attractions and repulsions of the charges in the atoms making up the molecules. Friction is also electromagnetic in origin, as is almost every force we see acting on the large scale. Tension? Due to chemical bonding. Drag force? Due to impacts of molecules on bodies moving through a fluid. Again, chemical, meaning electromagnetic.

Physicists would like to show somehow that all four fundamental forces are merely manifestations of one even more fundamental. So far, our best efforts haven't worked. In the 1970s, Glashow, Weinberg, and Salaam were able to show a fundamental link between the electromagnetic and weak force, but that's about it so far.

BIBLIOGRAPHY

[1] BBC News. Kepler 22-b: Earth-like planet confirmed. http://www.bbc.co.uk/news/science-environment-16040655.

[2] CERN. Antimatter FAQs. http://livefromcern.web.cern.ch/livefromcern/antimatter/faq.html.

[3] Crank Dot Net. Einstein was wrong. http://www.crank.net/einstein.html.

[4] The Extrasolar Planets Encyclopedia. Catalog, s.v. "Imaging." http://exoplanet.eu/catalog-imaging.php.

[5] The Extrasolar Planets Encyclopedia. Catalog, s.v. "Fomalhaut." http://exoplanet.eu/star.php?st=Fomalhaut.

[6] NASA. NASA discovers first earth-size planets beyond our solar system. News release. http://www.jpl.nasa.gov/news/news.cfm?release=2011-390.

[7] NASA. Hubble Space Telescope FAQs. http://www.spacetelescope.org/about/faq/.

[8] NASA. Kennedy Space Center FAQs. http://science.ksc.nasa.gov/pao/faq/faqanswers.htm.

[9] NASA. Nuclear Pulse Space Vehicle Survey. vol. I, Summary. Technical report. National Aeronautics and Space Administration, Huntsville, AL, September 1964.

[10] NASA. *Space Settlements: A Design Study*. Edited by Richard D. Johnson and Charles Holbrow. Technical Report NASA SP-413. National Aeronautics and Space Administration, Ames Research Center, 1975.

[11] National Space Society. Space settlement design contest 2009 results. http://www.nss.org/settlement/nasa/Contest/Results/2009/ASTEN.pdf.

[12] Space Exploration Technologies Corporation. Dragon. http://www.spacex.com/dragon.php/.

[13] Space Exploration Technologies Corporation. Falcon 1e pricing and performance. http://www.spacex.com/falcon1.php/#pricing_and_performance.

[14] Space.Com. Endeavor's shuttle launch delay comes with a large price tag. http://www.space.com/11525-space-shuttle-endeavour-launch-delay-cost.html.

[15] Television Tropes & Idioms. Recycled IN SPACE! http://tvtropes.org/pmwiki/pmwiki.php/Main/RecycledINSPACE.

[16] U.S. Census Bureau. Census death rate statistics. http://www.census.gov/compendia/statab/cats/transportation/motor_vehicle_accidents_and_fatalities.html.

[17] J. Ackeret. Zur Theorie der Raketen. *Helvetica Physica Acta*, 19:103–112, 1946.

[18] Douglas Adams. *The Hitchhiker's Guide to the Galaxy*. Harmony Books, New York, 1979.

[19] Douglas Adams. *The Restaurant at the End of the Universe*. Ballantine Books, New York, 2005.

[20] Fred C. Adams and Gregory Laughlin. A dying universe: The long-term fate and evolution of astrophysical objects. *Reviews of Modern Physics*, 69(2):337–372, 1997.

[21] Poul Anderson. *The Avatar*. Putnam, New York, 1978.

[22] Poul Anderson. *Tau Zero*. Doubleday, Garden City, NY, 1970.

[23] Poul Anderson. *The Earth Book of Stormgate*. Berkley, New York, 1978.

[24] P. K. Aravind. The physics of the space elevator. *American Journal of Physics*, 75(2):125–130, 2007.

[25] Svante Arrhenius and Joens Elias Fries. *The Destinies of the Stars*. G. P. Putnam's Sons, New York, 1918.

[26] Yuri P. Artutsanov. Skyhook: Old idea. *Science*, 158(3803):946–947, 1967.

[27] Yuri P. Artutsanov. Into space without rockets: A new idea for space launch. *Znaniye-Sila*, 7:25, 1969.

[28] Henri Arzeliès. *Relativistic Kinematics*. Pergamon Press, Oxford, 1966.

[29] W. L. Bade. Relativistic rocket theory. *American Journal of Physics*, 21:310–312, 1953.

[30] Arnold Barnett and Mary K. Higgins. Airline safety: The last decade. *Management Science*, 35(1):1–21, 1989.

[31] Roger R. Bate, Donald D. Mueller, and Jerry E. White. *Fundamentals of Astrodynamics*. Dover Publications, New York, 1971.

[32] Greg Bear. *The Forge of God*. Tor, New York, 1987.

[33] Greg Bear. *Moving Mars*. Tor, New York, 1993.

[34] Greg Bear. *Anvil of Stars*. Warner Books, New York, 1992.

[35] Edward Belbruno. *Fly Me to the Moon*. Princeton University Press, Princeton, NJ, 2007.

[36] John S. Bell. On the Einstein-Podolsky-Roden paradox. *Physics*, I, pp. 195–200, 1964.

[37] Gregory Benford and Larry Niven. *Bowl of Heaven*. Tom Doherty Associates, New York, 2012.

[38] Alfred Bester. *The Stars My Destination*. New American Library, New York, 1956.

[39] William J. Borucki et al. Kepler22b: A 2.4 Earth radius planet in the habitable zone of life of a sun-like star (preprint). http://arxiv.org/pdf/1112.1640v1, December 2011.

[40] Ray Bradbury. *The Martian Chronicles*. Doubleday, Garden City, NY, 1950.

[41] Kenneth Brower. *The Starship and the Canoe*. Holt, Rinehart & Winston, New York, 1978.

[42] Todd A. Brun. Computers with closed timelike curves can solve hard problems efficiently. *Foundations of Physics Letters*, 16(3):245–253, 2003.

[43] Edgar Rice Burroughs. *A Princess of Mars*. Doubleday, Garden City, NY, 1912.

[44] R. W. Bussard. Galactic matter and interstellar flight. *Astronomica Acta*, 6:97–111, 1960.

[45] Jim Butcher. *Proven Guilty*. Roc, New York, 2007.

[46] A. G. W. Cameron, ed. *Interstellar Communication: A Collection of Reprints and Original Contributions*. W. A. Benjamin, New York, 1963.

[47] Bradley W. Carroll and Dale A. Ostlie. *An Introduction to Modern Astrophysics*, 2nd ed. Addison-Wesley, San Francisco, CA, 2007.

[48] Hendrik G. B. Casimir. On the attraction between two perfectly conducting plates. *Proceedings of the Koninklijke Nederlandse Akademic van Wetenschappen*, 51:793–795, 1948.

[49] David Charbonneau, Timothy M. Brown, Robert W. Nayes, and Ronald L. Gilliland. Detection of an extrasolar planet atmosphere. *Astrophysical Journal*, 568:377–384, 2002.

[50] G. K. Chesterton. *The Napoleon of Notting Hill*. Dover Publications, New York, 1991.

[51] C. B. Clark. Experience in a non-inertial lab. *Physics Teacher*, 17:526, 1979.

[52] Arthur C. Clarke. *Childhood's End*. Harcourt, Brace & World, New York, 1953.

[53] Arthur C. Clarke. *The City and the Stars*. Harcourt, Brace, New York, 1956.

[54] Arthur C. Clarke. *Profiles of the Future*, rev. ed. Harper and Row, New York, 1973.

[55] Arthur C. Clarke. *The Promise of Space*. Harper & Row, New York, 1968.

[56] Arthur C. Clarke. *The Fountains of Paradise*. Harcourt Brace Jovanovich, New York, 1979.

[57] Arthur C. Clarke. *2010: Odyssey Two*. Ballantine Books, New York, 1982.

[58] Arthur C. Clarke and Stanley Kubrick. *2001: A Space Odyssey*. Hutchinson, London, 1968.

[59] Hal Clement. *Mission of Gravity*. Doubleday, Garden City, NY, 1954.

[60] Giuseppe Cocconi and Philip Morrison. Search for interstellar communications. *Nature*, 184:844–846, 1959.

[61] Paul J. Crutzen and Eugene F. Stoermer. The "anthropocene." *Global Change Newsletter*, 41:17–18, May 2000.

[62] Ingri D'Aulaire and Edgar Parin D'Aulaire. *D'Aulaire's Book of Norse Mythology*. New York Review of Books, New York, 1967.

[63] J. Dennis and L. Choate. Some problems with artificial gravity. *Physics Teacher*, 8:441, 1970.

[64] Jared Diamond. *The Third Chimpanzee: The Evolution and Future of the Human Animal*. HarperCollins, New York, 1992.

[65] Jared Diamond. *Guns, Germs, and Steel: The Fates of Human Societies*. W. W. Norton, New York, 1999.

[66] Jared M. Diamond. *Collapse: How Societies Choose to Fail or Succeed.* Viking, New York, 2005.

[67] Philip K. Dick. *Ubik.* Doubleday, Garden City, NY, 1969.

[68] Philip K. Dick. *Galactic Pot-Healer.* Vintage Books, New York, 1994.

[69] John D. Durand. Historical estimates of world population: An evaluation. *Population and Development Review*, 3(3):253–296, 1977.

[70] Damien Duvivier and Michel Wautelet. From the microworld to King Kong. *Physics Education*, 41:386–390, 2006.

[71] Freeman Dyson. Search for artificial stellar sources of infrared radiation. *Science*, 131(1341):1667–1668, 1960.

[72] Freeman Dyson. Interstellar transport. *Physics Today*, pp. 41–45, October 1968.

[73] Freeman Dyson. Time without end: Physics and biology in an open universe. *Reviews of Modern Physics*, 51(3):447–460, 1979.

[74] Freeman Dyson. The search for extraterrestrial technology. In *Selected Papers of Freeman Dyson*, 557–571. American Mathematical Society, 1996.

[75] George Dyson. *Project Orion: The True Story of the Atomic Spaceship.* Henry Holt, New York, 2002.

[76] Bradley C. Edwards. *The Space Elevator: Niac Phase 1 Report.* Technical report. NASA Institute for Advanced Concepts, Atlanta, GA, 2001.

[77] Bradley C. Edwards. *The Space Elevator: Niac Phase 2 Report.* Technical report. NASA Institute for Advanced Concepts, Atlanta, GA, 2003.

[78] A. Einstein, B. Podolsky, and N. Rosen. Can quantum-mechanical description of reality be considered complete? *Physical Review*, 47:777–780, 1935.

[79] Robert Erlich. Faster-than-light speeds, tachyons, and the possibility of tachyonic neutrinos. *American Journal of Physics*, 71(11):1109–1114, 2003.

[80] Gerald Feinberg. Possibility of faster-than-light particles. *Physical Review*, 159(5):1089–1105, 1967.

[81] Richard P. Feynman. *The Character of Physical Law.* MIT Press, Cambridge, MA, 1967.

[82] Richard P. Feynman, Anthony J. G. Hey, and Robin W. Allen. *The Feynman Lectures on Computation.* Addison-Wesley, New York, 1996.

[83] Richard P. Feynman and Ralph Leighton. *"What Do YOU Care What Other People Think?" Further Adventures of a Curious Character.* Norton, New York, 1988.

[84] Richard P. Feynman, Ralph Leighton, and Edward Hutchings. *"Surely You're Joking, Mr. Feynman!" Adventures of a Curious Character.* W. W. Norton, New York, 1985.

[85] Richard P. Feynman, Robert B. Leighton, and Matthew Sands. *The Feynman Lectures on Physics.* Addison-Wesley, New York, 1963.

[86] Martyn J. Fogg. *Terraforming: Engineering Planetary Environments.* Society of Automotive Engineers, Warrendale, PA, 1995.

[87] Robert L. Forward. Antiproton annihilation propulsion. Technical Report AD-A160 734. Air Force Rocket Propulsion Laboratory, Dayton, OH, 1985.

[88] Robert H. Frisbee. How to build an antimatter rocket for interstellar missions. In *Proceedings of the 39th AIAA/ASME/SAE/ASEE Joint Propulsion Conference and Exhibit*, AIAA-2003–4676, July 2003.

[89] Lo-Shu Fu. Teng Mu: A forgotten Chinese philosopher. *T'oung Pao*, 52:35–96, 1965.

[90] Galileo Galilei and Stillman Drake. *Dialogues Concerning Two New Sciences, Including Centers of Gravity & Force of Percussion*. University of Wisconsin Press, Madison, 1974.

[91] Hugo Gernsback. *Ralph 124C41+: A Romance of the Year 2660*. Stratford Co., Boston, 1925.

[92] David Gerrold. *A Matter for Men*. Simon & Schuster, New York, 1983.

[93] David Gerrold. *Jumping Off the Planet*. Tor, New York, 2000.

[94] David Gerrold. *Worlds of Wonder: How to Write Science Fiction & Fantasy*. Writer's Digest Books, Cincinnati, OH, 2001.

[95] William Gibson. *Neuromancer*. Ace Books, New York, 1984.

[96] William Gibson. *Mona Lisa Overdrive*. Bantam Books, Toronto, 1988.

[97] Samuel Glasstone and Philip J. Dugan. *The Effects of Nuclear Weapons*, 3rd ed. U.S. Department of Defense, Washington, DC, 1977.

[98] Robert Goddard. A method of reaching extreme altitudes. *Smithsonian Miscellaneous Collections*, 71(2):1–71, 1919.

[99] Herbert Goldstein. *Classical Mechanics*, 2nd ed. Addison-Wesley, Reading, MA, 1980.

[100] Guillermo Gonzalez. The galactic habitable zone. In *Astrophysics of Life*, ed. Mario Livio, I. Neill Reid, and William B. Sparks, 89–97. Cambridge University Press, Cambridge, 2005.

[101] David L. Goodstein, Richard P. Feynman, and Judith R. Goodstein. *Feynman's Lost Lecture: The Motion of Planets around the Sun*. Vintage Books, London, 1997.

[102] D. J. Griffiths, *Introduction to Quantum Mechanics*, 2nd ed. Pearson-Prentice Hall, Saddle River, NJ, 2000.

[103] Lev Grossman. *The Magicians*. Viking, New York, 2009.

[104] H. Chen et al. Relativistic positron creation using ultraintense short pulse lasers. *Physical Review Letters*, 102:105001-1–4, 2009.

[105] Thomas Hager. *The Alchemy of Air: A Jewish Genius, a Doomed Tycoon, and the Scientific Discovery That Fed the World but Fueled the Rise of Hitler*. Crown, New York, 2008.

[106] J. B. S. Haldane. *Possible Worlds and Other Essays*. Chatto & Windus, London, 1927.

[107] Joe W. Haldeman. *The Forever War*. St. Martin's Press, New York, 1974.

[108] Robert A. Heinlein. *Red Planet*, rev. ed. Ballantine, New York, 1947.

[109] Robert A. Heinlein. *Rocket Ship Galileo*. Scribner, New York, 1947.

[110] Robert A. Heinlein. *Space Cadet*. Ballantine Books, New York, 1948.

[111] Robert A. Heinlein. *Farmer in the Sky*. Scribner, New York, 1950.

[112] Robert A. Heinlein. *Starman Jones*. Ballantine Books, New York, 1953.

[113] Robert A. Heinlein. *Time for the Stars*. Scribner, New York, 1956.

[114] Robert A. Heinlein. *Have Space Suit—Will Travel*. Scribner, New York, 1958.

[115] Robert A. Heinlein. *Stranger in a Strange Land*. Putnam, New York, 1961.

[116] Robert A. Heinlein. *Glory Road*. Putnam, New York, 1963.

[117] Robert A. Heinlein. *Orphans of the Sky*. Putnam, New York, 1963.

[118] Robert A. Heinlein. *Podkayne of Mars*. Putnam, New York, 1963.

[119] Robert A. Heinlein. *The Moon Is a Harsh Mistress*. Putnam, New York, 1966.

[120] Robert A. Heinlein. *The Number of the Beast*. Fawcett, New York, 1980.

[121] Robert A. Heinlein. *Expanded Universe*. Baen, Riverdale, NY, 2003.

[122] Robert A. Heinlein and Clifford N. Geary. *Starman Jones*. Scribner, New York, 1953.

[123] T. A. Heppenheimer. On the infeasibility of interstellar ramjets. *Journal of the British Interplanetary Society*, 31:221–224, 1978.

[124] Walter Hohmann. *The Attainability of Heavenly Bodies*. NASA Technical Translation F-44, Washington, DC, 1961 (1925).

[125] Walter Hoppe, ed. *Biophysics*. Springer-Verlag, Berlin, 1983.

[126] M. King Hubbert. *Nuclear Energy and the Fossil Fuels*. Technical Report 95. Shell Development Co., Houston, TX, June 1956.

[127] M. King Hubbert. The world's evolving energy system. *American Journal of Physics*, 49(11):1007–1029, 1981.

[128] John D. Isaacs et al. Satellite elongation into a true "skyhook." *Science*, 151(3711):682–683, 1966.

[129] Daniel J. Jacob. *An Introduction to Atmospheric Chemistry*. Princeton University Press, Princeton, NJ, 1999.

[130] Jeff Hester et al. *21st Century Astronomy*, 2nd ed. W. W. Norton, New York, 2007.

[131] Chuck Jones. *Chuck Amuck: The Life and Times of an Animated Cartoonist*. Farrar, Straus and Giroux, New York, 1989.

[132] Eric M. Jones. *"Where Is Everybody?": An Account of Fermi's Question*. Technical Report LA-10311-MS. Los Alamos National Laboratory, Los Alamos, NM, 1985.

[133] W. Jones. Earnshaw's theorem and the stability of matter. *European Journal of Physics*, 1:85–88, 1980.

[134] J. F. Kasting, O. B. Toon, and J. B. Pollack. How climate evolved on the terrestrial planets. *Scientific American*, 256(2):90–97, 1988.

[135] James F. Kasting and David Catling. Evolution of a habitable planet. *Annual Reviews of Astronomy and Astrophysics*, 41:429–463, 2003.

[136] James F. Kasting, Daniel P. Whitmire, and Ray T. Reynolds. Habitable zones around main sequence stars. *Icarus*, 101:108–128, 1993.

[137] N. Y. Kiang. The color of plants on other worlds. *Scientific American*, 298(4):48–55, 2008.

[138] N. Y. Kiang et al. Spectral signatures of photosynthesis. I. Review of earth organisms. *Astrobiology*, 7:222–251, 2007.

[139] N. Y. Kiang et al. Spectral signatures of photosynthesis. II. Coevolution with other stars and the atmosphere on extrasolar worlds. *Astrobiology*, 7:252–274, 2007.

[140] Max Kleiber. *The Fire of Life: An Introduction to Animal Energetics*. John Wiley & Sons, New York, 1961.

[141] Elizabeth Kolbert. *Field Notes from a Catastrophe: Man, Nature, and Climate Change*. Bloomsbury, New York, 2006.

[142] D. G. Korycansky, G. Laughlin, and F. C. Adams. Astronomical engineering: A strategy for modifying planetary orbits. *Astrophysics and Space Science*, 275:349–366, 2001.

[143] Rolf Landauer. Irreversability and heat generation in the computing process. *IBM Journal of Research Development*, 5:183, 1961.

[144] J. Laskar and P. Robutel. The chaotic obliquity of the planets. *Nature*, 361:608–612, 1993.

[145] Ursula K. Le Guin. *The Left Hand of Darkness*. Harper & Row, New York, 1969.

[146] Alfred Leick. *GPS Satellite Surveying*, 2nd ed. John Wiley & Sons, New York, 1995.

[147] Stanislaw Lem. *His Master's Voice*. Harcourt Brace Jovanovich, San Diego, CA, 1968.

[148] Stanislaw Lem. *Solaris*. Harcourt Brace Jovanovich, San Diego, CA, 1970.

[149] Stanislaw Lem. *A Perfect Vacuum*. Harcourt Brace Jovanovich, New York, 1979.

[150] Stanislaw Lem. *Microworlds*. Harcourt, Brace & Co., San Diego, CA, 1984.

[151] Stanislaw Lem. *Fiasco*. Harcourt Brace Jovanovich, New York, 1988.

[152] Madeleine L'Engle. *A Wrinkle in Time*. Farrar, Straus and Giroux, New York, 1962.

[153] Mario Livio. *The Accelerating Universe: Infinite Expansion, the Cosmological Constant, and the Beauty of the Cosmos*. Wiley, New York, 2000.

[154] Edward N. Lorenz. Deterministic nonperiodic flow. *Journal of the Atmospheric Sciences*, 20:130–141, 1963.

[155] Percival Lowell. *Mars*. Houghton Mifflin, Boston, 1895.

[156] John Maddox, Poul Anderson, Eugene A. Sloan, and Freeman Dyson. Artificial biospheres. *Science*, 132:250–253, 1960.

[157] Michael E. Mann and Lee R. Kump. *Dire Predictions: Understanding Global Warming*. DK Publishing, London, 2008.

[158] Anthony R. Martin. Some limitations of the interstellar ramjet. *Spaceflight*, 14:21–25, 1972.

[159] Anthony R. Martin. Magnetic intake limitations on interstellar ramships. *Astronomica Acta*, 18:1–10, 1973.

[160] John H. Mauldin. *Prospects for Interstellar Travel*. Science and Technology Series 80. American Astronautical Society, Springfield, VA, 1992.

[161] James Clerk Maxwell. *On the Stability of the Motion of Saturn's Rings*. Macmillan, London, 1856.

[162] Colin McInnes. Non-linear dynamics of ring world systems. *Journal of the British Interplanetary Society*, 56:308–313, 2003.

[163] Christopher P. McKay and Wanda L. Davis. Duration of liquid water habitats on early Mars. *Icarus*, 90:214–221, 1991.

[164] John McPhee. *The Curve of Binding Energy*. Farrar, Straus and Giroux, New York, 1974.

[165] N. David Mermin. *It's about Time: Understanding Einstein's Relativity*. Princeton University Press, Princeton, NJ, 2005.

[166] N. David Mermin. *Quantum Computer Science: An Introduction*. Cambridge University Press, Cambridge, 2007.

[167] Robert A. Mole. Terraforming Mars with four war surplus bombs. *Journal of the British Interplanetary Society*, 48(7):321–324, 1995.

[168] Michael S. Morris and Kip Thorne. Wormholes in spacetime and their use for interstellar travel: A tool for teaching general relativity. *American Journal of Physics*, 56(5):395–412, 1988.

[169] Michael S. Morris, Kip Thorne, and Ulvi Yurtsaever. Wormholes, time machines and the weak energy condition. *Physical Review Letters*, 61(13):1446–1449, 1988.

[170] D. J. Mullan. Why is the sun so large? *American Journal of Physics*, 74(1):10–13, 2006.

[171] Thomas Muller. Visual appearance of a Morris-Thorne wormhole. *American Journal of Physics*, 72(8):1045–1050, 2004.

[172] Randall Munroe. xkcd: Quantum teleportation. http://xkcd.com/465/.

[173] Paul J. Nahin. *The Logician and the Engineer: How George Boole and Claude Shannon Created the Information Age*. Princeton University Press, Princeton, NJ, 2012.

[174] Paul J. Nahin. *Time Machines: Time Travel in Physics, Metaphysics and Science Fiction*. American Institute of Physics Press, New York, 1993.

[175] Larry Niven. *A Gift from Earth*. Walker, New York, 1968.

[176] Larry Niven. *Neutron Star*. Ballantine Books, New York, 1968.

[177] Larry Niven. *Ringworld*. Holt, Rinehart & Winston, New York, 1970.

[178] Larry Niven. *All the Myriad Ways*. Ballantine Books, New York, 1971.

[179] Larry Niven. *Protector*. Ballantine Books, New York, 1973.

[180] Larry Niven. *A Hole in Space*. Ballantine Books, New York, 1974.

[181] Larry Niven. *Tales of Known Space*. Ballantine Books, New York, 1975.

[182] Larry Niven. *A World Out of Time*. Holt, Rinehart & Winston, New York, 1976.

[183] Larry Niven. *The Ringworld Engineers*. Holt, Rinehart & Winston, New York, 1980.

[184] Larry Niven. *Convergent Series*. Del Rey, New York, 1986.

[185] Larry Niven and Brenda Cooper. *Building Harlequin's Moon*. Tor, New York, 2005.

[186] Larry Niven and Jerry Pournelle. *The Mote in God's Eye*. Simon & Schuster, New York, 1974.

[187] Larry Niven and Jerry Pournelle. *Footfall*. Del Rey, New York, 1985.

[188] Gerard K. O'Neill. The colonization of space. *Physics Today*, 27:32–40, 1974.

[189] Gerard K. O'Neill. *The High Frontier: Human Colonies in Space*. William Morrow, New York, 1977.

[190] J. Pearson. The orbital tower: A spacecraft launcher using the earth's rotational energy. *Acta Astronautica*, 2:785–799, 1975.

[191] C. J. Pennycuick. *Bird Flight Performance: A Practical Calculation Manual*. Oxford University Press, Oxford, 1989.

[192] Henri Poincaré. *The Foundations of Science*. Science Press, New York, 1929.

[193] J. B. Pollack, J. F. Kasting, S. M. Richardson, and K. Poliakoff. The case for a warm, wet climate on early Mars. *Icarus*, 71:203–224, 1987.

[194] Michael Pollan. *The Omnivore's Dilemma: A Natural History of Four Meals*. Penguin Press, New York, 2006.

[195] Tim Powers. *On Stranger Tides*. Ace Books, New York, 1987.

[196] Tim Powers. *Last Call*. William Morrow, New York, 1992.

[197] Richard Rhodes. *The Making of the Atomic Bomb*. Simon & Schuster, New York, 1986.

[198] Wolfgang Rindler. *Special Relativity*, 2nd ed. Oliver and Boyd, London, 1966.

[199] Kim Stanley Robinson. *Forty Signs of Rain*. Bantam Books, New York, 2004.

[200] A. Robock et al. Climatic consequences of regional nuclear conflicts. *Atmospheric Chemistry and Physics*, 7(8):2003–2012, 2007.

[201] J. K. Rowling. *Harry Potter and the Sorcerer's Stone*. A. A. Levine Books, New York, 1998.

[202] J. K. Rowling. *Harry Potter and the Prisoner of Azkaban*. Arthur A. Levine Books, New York, 1999.

[203] J. K. Rowling. *Harry Potter and the Goblet of Fire*. Arthur A. Levine Books, New York, 2000.

[204] J. K. Rowling. *Harry Potter and the Order of the Phoenix*. Arthur A. Levine Books, New York, 2003.

[205] J. K. Rowling. *Harry Potter and the Half-Blood Prince*. Arthur A. Levine Books, New York, 2005.

[206] J. K. Rowling. *Harry Potter and the Deathly Hallows*. Arthur A. Levine Books, New York, 2007.

[207] R. S. Ruoff, S. Quian, and W. K. Liu. Mechanical properties of carbon nanotubes. *Comptes-Rendus Physique*, 4:993–1008, 2003.

[208] Carl Sagan. Planetary engineering on Mars. *Icarus*, 20:513–514, 1973.

[209] Carl Sagan. *Contact*. Simon & Schuster, New York, 1985.

[210] J. L. Sanders. Advanced post-*Saturn* Earth launch vehicle study: Executive summary report. Technical Report TM X-53200. National Aeronautics and Space Administration, 1965.

[211] Hans Joachin Schellnhuber, Paul J. Crutzen, William C. Clark, and Julian Hunt. Earth systems analysis and sustainability. *Environment*, 47(8):11–25, 2005.

[212] A. G. Schmidt. Coriolis acceleration and conservation of angular momentum. *American Journal of Physics*, 54:755, 1986.

[213] Sara Seager and Drake Deming. Exoplanet atmospheres. *Annual Review of Astronomy and Astrophysics*, 48(1):631–672, 2010.

[214] Antigona Segura et al. Ozone concentrations and ultraviolet fluxes on Earth-like planets around other stars. *Astrobiology*, 3:689–708, 2003.

[215] Claude Semay and Bernard Silvestre-Brac. Equation of motion of an interstellar Bussard ramjet with radiation loss. *Acta Astronautica*, 61:817–822, 2007.

[216] Charles Sheffield. *The Web between the Worlds*. Ace Books, New York, 1979.

[217] Charles Sheffield. *The Compleat McAndrew*. Baen Books, Riverdale, NY, 2000.

[218] Michael Shermer. Why ET hasn't called. *Scientific American*, August 2002.

[219] Daniel Simmons. *Hyperion*. Doubleday, New York, 1989.

[220] Daniel Simmons. *The Fall of Hyperion*. Doubleday, New York, 1990.

[221] E. E. Smith. *First Lensman*. Fantasy Press, Reading, PA, 1950.

[222] George O. Smith. *The Complete Venus Equilateral*. Ballantine Books, New York, 1976.

[223] S. Solomon, D. Qin, M. Manning, Z. Chen, M. Marquis, K. B. Averyt, M. Tignor, and H. L. Miller, ed. *IPCC, 2007: Climate Change 2007: The Physical Science Basis*. Cambridge University Press, Cambridge, 2007.

[224] Walter S. Stahl. Scaling of respiratory variables in mammals. *Journal of Applied Physiology*, 22(3):453–460, 1967.

[225] Olaf Stapledon. *Last and First Men & Star Maker*. Dover Publications, New York, 1968.

[226] John Steinbeck. *The Grapes of Wrath*. Viking Press, New York, 1939.

[227] Arthur Stinner and John Begoray. Journey to Mars: The physics of traveling to the red planet. *Physics Education*, 40(1):35–45, 2005.

[228] S. M. Stirling. *The Stone Dogs*. Baen Books, Wake Forest, NC, 1990.

[229] S. M. Stirling. *The Sky People*. Tor, New York, 2006.

[230] Arkady Strugatsky and Boris Strugatsky. *Roadside Picnic & Tale of the Troika*. Macmillan, New York, 1977.

[231] Yuri M. Svirezhev. Non-linearities in mathematical ecology: Phenomena and models. *Ecological Modeling*, 216(2):89–101, 2008.

[232] Victor Szebehely. *Theory of Orbits: The Restricted Problem of Three Bodies*. Academic Press, New York, 1967.

[233] T. Coan et al. A compact apparatus for muon lifetime measurement and time dilation demonstration in the undergraduate laboratory. *American Journal of Physics*, 74(2):161–162, 2006.

[234] Edwin F. Taylor and John Archibald Wheeler. *Spacetime Physics: Introduction to Special Relativity*, 2nd ed. W. H. Freeman, New York, 1992.

[235] Edwin F. Taylor and John Archibald Wheeler. *Exploring Black Holes: Introduction to General Relativity*. Addison-Wesley-Longman, Boston, 2000.

[236] Kip S. Thorne. *Black Holes and Time Warps: Einstein's Outrageous Legacy*. W. W. Norton, New York, 1994.

[237] Frank J. Tipler. Rotating cylinders and the possibility of global causality violation. *Physical Review D*, 9(8):2203–2206, 1974.

[238] Akira Tomizuka. Estimation of the power of greenhouse gases on the basis of absorption spectra. *American Journal of Physics*, 78(4):359–366, 2010.

[239] Virginia Trimble. Cosmology: Man's place in the universe, a deconstruction. *American Journal of Physics*, 70(12):1175–1183, 2002.

[240] S. M. Ulam. *Analogies between Analogies: The Los Alamos Reports of S. M. Ulam*. University of California Press, Berkeley, 1990.

[241] Jules Verne. *De la Terre à la Lune*. J. Hetzel et Cie. Paris, 1866.

[242] Matt Visser. *Lorentzian Wormholes*. American Institute of Physics, Woodbury, NY, 1996.

[243] Steven Vogel. *Comparative Biomechanics: Life's Physical World.* Princeton University Press, Princeton, NJ, 2003.

[244] Sarah Vowell. *Assassination Vacation.* Simon & Schuster, New York, 2005.

[245] David Walker and Richard Walker. *Energy, Plants and Man,* 2nd ed. Oxygraphics, 1992.

[246] Jearl Walker. *Fundamentals of Physics,* 8th extended ed. John Wiley & Sons, New York, 2008.

[247] Peter D. Ward and Donald Brownlee. *Rare Earth.* Copernicus, New York, 2000.

[248] H. G. Wells. *The War of the Worlds.* Harper & Brothers, New York, 1898.

[249] H. G. Wells. *The First Men in the Moon.* G. Newnes, London, 1901.

[250] John Archibald Wheeler and Richard Phillips Feynman. Interaction with the absorber as the mechanism of radiation. *Reviews of Modern Physics,* 17:157–181, 1945.

[251] T. H. White. *The Once and Future King.* Collins, London, 1958.

[252] Clifford M. Will. *Was Einstein Right? Putting General Relativity to the Test.* Basic Books, New York, 1986.

[253] Gene Wolfe. *The Fifth Head of Cerberus.* Scribner, New York, 1972.

[254] Gene Wolfe. *The Citadel of the Autarch.* Vol. 4 of *Book of the New Sun.* Timescape Books, New York, 1983.

[255] W. K. Wootters and W. H. Zurek. A single quantum cannot be cloned. *Nature,* 299:802–803, 1982.

[256] Robert M. Zubrin and Christopher P. McKay. Technological requirements of terraforming Mars. In *Bringing Mars to Life,* ed. R. L. S. Taylor. British Interplanetary Society, 1994.

INDEX